Practical Vacuum Systems

Practical Vacuum Systems

Rolland Rutledge LaPelle

Senior Member, American Vacuum Society
Formerly Chairman, Vacuum Gauge Calibration Committee,
Committee E-21 (Space Simulation Standards),
American Society of Testing and Materials
Past Chairman, Pacific Northwest Section, American Vacuum Society
Past President, Astronomical League
The Boeing Company Senior Research Engineer, Ret.

McGRAW-HILL BOOK COMPANY

New York St. Louis San Francisco Düsseldorf Johannesburg
Kuala Lumpur London Mexico Montreal New Delhi
Panama Rio de Janeiro Singapore Sydney Toronto

Library of Congress Cataloging in Publication Data

LaPelle, Rolland Rutledge
Practical vacuum systems.

Bibliography: p.
1. Vacuum technology. I. Title.
TJ940.L36 621.5′5 72-4518
ISBN 0-07-036355-2

1234567890 KPKP 765432

The editors for this book were Tyler G. Hicks, Don A. Douglas, and Karen Kesti, the designer was Naomi Auerbach, and its production was supervised by George E. Oechsner. It was set in Caledonia by The Maple Press Company.

It was printed and bound by The Kingsport Press.

To Sally, without whose patience, persistent urging, and continuous encouragement, this book would never have been completed

Contents

Preface ix

1. The Nature of Vacuum 1
2. Typical Vacuum Systems 4
3. Materials of Construction 8
4. Mechanical Vacuum Pumps 23
5. Vapor-type Pumps 32
6. Ionic and Sublimation Pumps 46
7. Trouble-shooting, Cleaning, and Repairing Vacuum Pumps 56
8. Cryogenics in Vacuum Systems 67
9. Vacuum Gauges 78
10. Vacuum Gas Analyzers 94
11. Vacuum Gauge Calibration 100
12. The Vacuum Vessel 117
13. Welding for High Vacuum 124
14. Closures 132
15. Accessories 141

16. Finishing, Cleaning, and Backfilling Systems 157
17. Leak-detection Techniques 162
18. Theory of Gases 171
19. Flow of Gases 179
20. Pumping Calculations 190
21. Uses of Vacuum Systems 202
Appendix A. Table of Conversion Factors for Pumping Systems 225
Appendix B. Bibliography 229
Index 231

Preface

This book is written to provide necessary information for technicians, engineering aids, and engineers, who are required to build, operate, or rework vacuum systems in connection with production, test, or experimental equipment. It covers primarily the practical applications of the vacuum art, and it will employ basic vacuum theory only to the extent necessary to an understanding of the phenomena involved and to the selection of suitable formulas for computing conductance, pumpdown time, and pumping requirements.

Since the book has been designed for students without previous courses in calculus or college level physics, the formulas given have been limited to simple algebraic equations. As a result, derivation of the equations has not been possible. However, references are given to more complete treatments elsewhere. The scales for dimensions, pressure, flow, and temperature are those in common use, with no attempt at consistency by adhering to metric units throughout. The units in the book are those commonly used in industry (though not necessarily in research laboratories) although this does involve a mixture of several systems of mensuration. Manufacture of equipment in industrial shops is almost always carried out in English units, and alternative formulas are given for these units. The factors necessary to convert from one scale to another are given in convenient form in Appendix A.

The terminology employed for various items in the vacuum field is that recommended in the "Glossary of Terms Used in Vacuum Technology" issued by the Committee on Standards of the American Vacuum Society and/or by the Nomenclature Subcommittee of Committee E-21 on Space Simulation of the American Society for Testing and Materials. These definitions are the ones in general use in the United States and are not necessarily in agreement with international standards having to do with terminology in physics, chemistry, and other fields. Our intent has been to speak in terms in common use in American research, engineering, and manufacturing organizations.

This book grew out of a number of courses given at the Boeing Company and in conjunction with the extension work of the Seattle Public School System. The material presented has been selected to be of maximum usefulness to those taking the courses. For this reason material has been included covering not only vacuum systems, their fabrication and repair, but a variety of accessory equipment associated with the vacuum systems, including cryogenics.

In a field such as ultrahigh vacuum, which is changing with extreme rapidity, it is probable that the specific devices described herein will be out of date within a few years from the date of writing. The student is therefore urged to make use of the latest data, as available in manufacturers' literature and in publications in the vacuum field, in order to keep abreast of the advancing technology. The transactions of the annual Symposia of the American Vacuum Society, as published in the *Journal of Vacuum Science and Technology,* will be found to be most helpful in this way.

Our hope is that this book may be valuable to those working in the vacuum field who have need to design, construct, or operate vacuum systems, but whose primary interest is in the production, research, or experimental work to be done within the system rather than with the theory of vacuum itself.

Rolland R. LaPelle

Practical Vacuum Systems

chapter 1

The Nature of Vacuum

1.1 *Definitions*

Technically, a vacuum system is any system in which the pressure is held at a value below atmospheric pressure at any given time. Vacuum systems of a relatively crude sort are used in many industrial processes, including concentration of syrup to make sugar, concentration and dehydration of a variety of food products, the production of biologicals and pharmaceuticals, the vacuum cooling of perishable crops such as lettuce, and a great number of processes involving removal of water vapor from mixtures that are cooled through the effects of evaporation. Vacuum is also essential to the efficient operation of all large steam-turbine generators and other prime movers. Such vacuums as these are generally produced through the use of surface condensers or relatively crude evacuating pumps, blowers, or steam ejectors. The degree of vacuum is frequently measured in terms of inches of mercury below atmosphere. Thus we speak of a vacuum of 28 inches in a power-plant condenser system, or 29 inches in certain other processes.

In this book, however, we are concerned with vacuums of a much higher degree than these. In vacuum systems, the absolute pressure is frequently measured in terms of millimeters of mercury where the standard atmosphere is taken as 760 millimeters of mercury at a temperature of 0°C. In this book, we shall be dealing with pressures from approximately 1 millimeter of mercury down to 10^{-9} millimeters of mercury or below. The term "millimeters of mercury" being somewhat lengthy either to express or write, the American Vacuum Society and the International Vacuum Society have standardized it as torr, where one millimeter of mercury is taken as equal to one torr. Thus a pressure of 1×10^{-5} millimeters of mercury is usually written as 1×10^{-5} torr, abbreviated as T. While the formal definition involves slightly different concepts, for all practical purposes one torr may be considered to equal one millimeter of mercury.

There are many processes involving pressures of 1×10^{-3} torr or more. For convenience, 1×10^{-3} torr is called 1 micron. See Appendix A for units of pressure.

1.2 *Degrees of Vacuum*

Many schemes of nomenclature have been proposed for naming vacuum systems having various capabilities in terms of pressure. Here we shall adhere to the recommendations of the American Vacuum Society Standards Committee as follows:

High vacuum.................	1×10^{-3} to 1×10^{-5} torr
Very high vacuum.............	1×10^{-6} to 1×10^{-8} torr
Ultrahigh vacuum.............	1×10^{-9} torr and below

The significance of these classifications becomes evident when it is realized that the cost of a vacuum system varies inversely as some power of the pressure intended. Thus it is a serious waste of money to design a system capable of attaining a much lower pressure than that for which it is intended.

However, it is an equally serious waste of money to design inadequately for the final pressure required. This is especially true since the cost of troubleshooting, reworking, and rebuilding is far greater than the cost of having specified a better construction in the first place, and in addition a large amount of time is wasted during the rework period.

We shall therefore attempt, in setting forth the constructional methods, to indicate the pressure limits at which each variety of materials and accessories may be successfully used, so that a proper choice may be made in the full knowledge of the consequences thereof.

The use to which a vacuum system will be put will influence in a major way the pumping methods, details of construction, and the vacuum level required. In addition, in most systems special requirements exist for manipulators, mask-changing devices, cryogenic surfaces for space simulation, solar simulation, and many other details which affect the vacuum system proper. We shall outline these requirements in more detail in later chapters.

1.3 *Expansion Ratios*

In a later chapter we shall go in detail into the nature of the gas laws. However, as an introduction it is instructive to consider the expansion ratios that are entailed by the achievement of the vacuum environment. Basing our understanding on the simple gas law that $P_1 \times V_1$ must equal $P_2 \times V_2$, consider the following table, based upon the expansion of a volume of 1 cubic foot of gas at various pressures.

Pressure, torr	*Volume, cu ft*
760 (760 mm or 1 atm)	1
1 (1 mm)	760
1×10^{-1} (100 microns)	7,600
1×10^{-2} (10 microns)	76,000
1×10^{-3} (1 micron)	760,000
1×10^{-4}	7,600,000
1×10^{-5}	76,000,000

It becomes immediately apparent that if the continuous inleak of gas is 1 cubic foot per minute at atmospheric pressure, the pumping speed required to reach a pressure of 1×10^{-5} torr is so large as to be thoroughly impractical.

chapter 2

Typical Vacuum Systems

2.1 General

Any vacuum system must have certain basic parts which are common to all systems regardless of pumping methods employed. These are as follows:

1. A gas-tight vacuum vessel with gas-tight closures where entrance can be made at some phase of the operating cycle.
2. A rough-pumping system which will reduce the pressure from atmospheric to a level where low-pressure systems can be used.
3. A fine-pumping system which is capable of reaching the ultimate pressure the system must attain with sufficient pumping speed to handle the outgassing which results from work carried out within the vessel.
4. A system of vacuum gauges and readouts to enable the pressure to be measured both during the roughing stage and during the fine-vacuum stage.

2.2 Diffusion-pumped Systems

The most common vacuum systems for a wide variety of applications are systems using a vapor diffusion pump as the fine-pumping source. A typical system of this type is shown in Fig. 2.1.

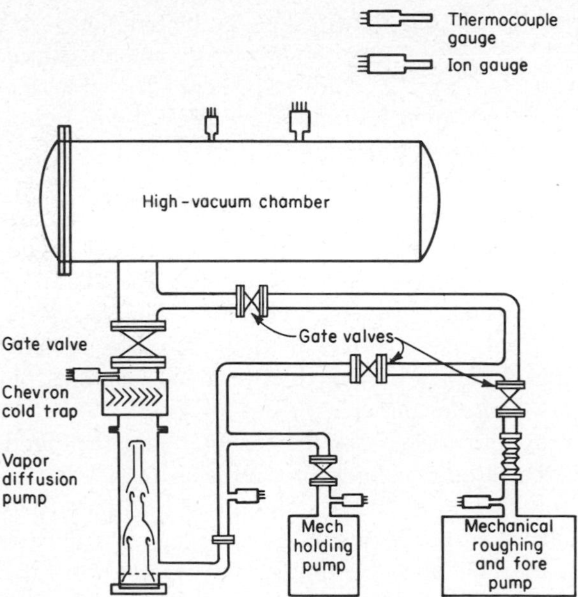

Fig. 2.1 Schematic of a typical high-vacuum system.

In such a system, the rough pumping is done by mechanical pumps of either the vane or rotary-piston type, capable of exhausting directly to the atmosphere. Such pumps reduce the system pressure sufficiently to allow the diffusion pump to operate. In small systems, it is conventional to utilize the roughing pump as a backing pump for the diffusion pump after the latter is started. In very large systems, there are frequently several roughing pumps which are much larger than required for backing the diffusion pumps. In such a system, it is conventional to isolate the roughing pumps at the end of the roughing cycle and use a separate pump or one of the roughing pumps as backing pumps. This reduces the electrical load during the fine-pumping portion of the cycle when only a small fraction of the air remains in the system.

Diffusion pumps have no moving parts and achieve their pumping action by boiling a fluid (oil) in the bottom of the pump. The resulting stream of oil vapor travels up through a center chimney and out as a series of jets at high velocity toward the diffusion-pump wall. These dense streams of rapidly moving oil vapor collide with molecules of the gas being pumped and push them downward toward the outlet of the diffusion pump. They are removed by the mechanical backing pump and exhausted to the atmosphere. The oil vapor condenses and returns to the boiler.

Above the diffusion pump, a "cold trap" is generally employed to prevent the migration of oil molecules into the working vacuum chamber. The temperature required in the cold trap depends upon the vacuum level to be obtained and upon the vapor pressure of the pump fluid used. Water- or Freon-cooled traps may be used with modern diffusion-pump fluids for most vacuum processes. Liquid-nitrogen-cooled traps of proper design can be built which prevent most of the backstreaming from reaching the work zone. Except in ultrahigh-vacuum systems, the cold trap is generally separate from the chamber proper by means of a gate valve of large size. This arrangement allows the pump to be isolated from the chamber without waiting for the pumping fluid to cool. Rapid backfilling on these "valved" systems allows quick changing of work in the chamber.

Reference to the diagram indicates a small holding pump arranged so that it can be cut into the circuit to back the diffusion pump during periods when the main roughing and backing pump is being used to reevacuate the chamber after a specimen has been changed. This is necessary since the pump foreline pressure should not exceed a critical level with the boiler operating.

Further reference to the diagram will indicate the presence of a number of vacuum gauges which enable the blankoff pressure of each rough-pumping unit to be independently measured. Such an array of gauges is not essential to the operation of the system but is a tremendous aid in troubleshooting. Malfunctions in the system can be checked by examining the vacuum level of each individual pump and the proper corrective measures more easily applied. Inexpensive thermocouple-type gauges are adequate for all of these points. A common readout system can be used for troubleshooting by connecting or switching it to the gauge being used. The main chamber itself must be provided with an ion gauge adequate in design to read properly the pressure to be attained in the chamber, since vacuum systems normally reach pressure levels too low to be read by the simple thermocouple-type gauges.

2.3 *Ion-pumped Systems*

Many systems today are being designed to utilize the Penning discharge pump, commonly termed an "ion pump." These pumps secure a pumping effect not by ejecting the gas from the sealed volume, but by immobilizing it by chemical combination or by burial in the anode and cathode structures of the pump itself. Thus they operate without backing during their normal operating cycle.

Some variety of roughing system is required in order to reduce the pressure to the point at which an ion pump will start. For most varieties of pumps currently on the market, the required starting pressure is between 3×10^{-3} and 5×10^{-4} torr. In large chambers, this requires a sophisticated roughing system. Ordinary mechanical pumps, even of the two-stage variety, do not quickly reach this pressure with a large gas load.

The pumping systems employed for roughing ion-pumped chambers are usually of one of two types. The first of these makes use of the ability of certain molecular sieve materials (artificial zeolites) to pump or adsorb the gases of the air when cooled by liquid nitrogen to approximately 78°K. Such systems generally make use of two or more "sorption pumps," as they are commercially known, each provided with a cooling dewar for the reception of the liquid nitrogen. The pumps are used in sequence to speed the system pumpdown to a point where the ion pump will start. At this point the ion pump is energized, the sorption pumps remain in operation, and in due time ion pumping commences. The sorption pumps are then closed off from the system, generally by means of a bakeable valve employing a metal seat. The system is now completely isolated from the atmosphere. A mechanical pump may be used to reduce the amount of sorption pumping required. When this is done, it is very important that a trap be employed between the mechanical pump and the manifold to prevent pump oil from getting into the chamber. The major advantage of sorption pumping is absolute cleanliness.

As an alternate to the sorption pumps just described, many ion-pumped systems employ roughing systems consisting of mechanical pumps combined with Roots-type blowers and liquid-nitrogen traps. Such systems are considerably more expensive than the simple roughing systems used on diffusion-pumped systems, but they need not be restricted to roughing a single chamber. It is convenient to mount the roughing-pump system on a wheeled cart which may be connected to any one of two or more ion-pumped systems, being employed when starting is desired and disconnected at the end of the roughing period. Thus the relatively expensive roughing-pump system required for ion pumps may be spread over several chambers and the cost per chamber brought down to a reasonable level.

An alternative to using a Roots blower roughing system, as described above, is to use a turbine or molecular-drag-type pump. Unfortunately, molecular-drag pumps are quite expensive and, for that reason, are not widely used.

chapter 3

Materials of Construction

3.1 *The Importance of Vapor-pressure Information*

All materials have characteristic vapor pressures which vary with temperature. If a vacuum system is to be capable of achieving the low pressure ranges in which we are interested, no material exposed to vacuum can have a vapor pressure such that it will volatilize at the expected operating temperatures and pressures. It is perhaps obvious that the materials which we readily recognize as having a low boiling point—such as water, grease, and ordinary oils—could be expected to give trouble in a high-vacuum system. What is not so obvious, however, is that materials which we do not ordinarily think of as being readily volatilized will also give off vapors which will impair the integrity of vacuum systems. Such materials as wood, leather, cloth, paper, asbestos, and some forms of rubber and plastic break down or outgas rather readily at room temperatures when subjected to high vacuum, giving off large quantities of vapors which are released slowly but continuously for a long period of time, sometimes leading to ultimate breakdown and degradation of the materials and, of course, adversely affecting the system pressure so that hard vacuums cannot be achieved. Materials of construction for vacuum systems must therefore be carefully selected in order to prevent this volatilization of the various gaseous components of such materials.

3.2 *Vapor Pressure of Organic Materials*

Table 3.1 shows a number of vacuum materials which are useful under varying conditions. These include the greases and waxes used for seals in glass systems and for lubrication of gaskets and O rings. It will be noticed that while some of these materials have very low vapor pressures, they must be used cold or relatively cool because they melt at a rather moderate temperature and become useless at this point. In addition, at extremely low temperatures such as those achieved in cold traps and other cryogenic equipment, these materials become hard and brittle and lose their elasticity, and therefore their sealing ability. They must therefore be used in vacuum systems at approximately room temperatures and carefully maintained at these temperatures during operation.

3.3 *Vapor Pressures of Pump Fluids and Plastics*

Table 3.2 shows the characteristics of a variety of plastic materials used for gaskets, insulation, etc. In general, any parts exposed to high vacuum should have vapor pressures at least as low as the pressure to be attained in the system. This severely limits the possible choices for the lower-pressure vacuum systems.

One of the points to notice here is that many of the insulators used in electrical wire and cable manufacture have relatively high vapor pressures. It is therefore inadvisable to use standard insulated wires inside vacuum systems. If the vacuum system is to be pumped to extremely low pressure, the wire must be either originally insulated by means of Teflon insulation or stripped of its as-received insulation and reinsulated. Teflon is satisfactory as an insulating material in vacuum systems down to pressures corresponding to its vapor pressure. However, for ultrahigh-vacuum systems the system pressure is lower than the vapor pressure of Teflon. For these systems, the insulation must be made of dense, highly purified ceramics or glass or quartz tubing, which have negligible vapor pressures and are relatively easily cleaned and outgassed. A failure to take into account the outgassing of the insulation

TABLE 3.1 Characteristics of Greases, Waxes, and Cements

Trade or chemical name	Supplier	Melting point, °C	Vapor pressure at 20°C, torr	General application notes
Apiezon grease L	J.G.B.	47	10^{-10}–10^{-11}; 10^{-3} at 300°C	Maximum temperature 30°C
Apiezon grease M	J.G.B.	44	10^{-7}–10^{-8}; 10^{-3} at 200°C	More viscous than Apiezon L; maximum temperature 30°C
Apiezon grease N	J.G.B.	43	10^{-8}–10^{-9}; 10^{-3} at 200°C	Maximum temperature 30°C
Apiezon Q	J.G.B.	45	10^{-4}	Maximum temperature 30°C; soft wax
Apiezon wax W-40	J.G.B.	45	10^{-3}	Maximum temperature 30°C; for joints subject to vibration; medium-soft wax
Apiezon wax W-100	J.G.B.	55	10^{-3}	Maximum temperature 50°C; for joints subject to vibration; medium-hard wax
Apiezon wax W	J.G.B.	85	10^{-3} at 180°C	S.P. = 60–70°C; maximum temperature 80°C; for permanent joints; soluble in xylene
Apiezon grease T	J.G.B.	25	10^{-8}	Maximum temperature 110°C
Apiezon oil J	J.G.B.	.....	10^{-7}–10^{-9}; 10^{-3} at 250°C	Moderately viscous oil
Apiezon oil K	J.G.B.	.....	10^{-9}–10^{-10}; 10^{-3} at 300°C	Exceedingly viscous oil
Celvacene, light	C.V.C.	90	10^{-6}	Pale yellow transparent grease
Celvacene, medium	C.V.C.	120	$<10^{-6}$	Yellow to brownish transparent grease
Calvacene, heavy	C.V.C.	130	$<10^{-6}$	Dark yellow to reddish brown transparent grease; soluble in CH_3Cl and acetone
DeKhotinsky cement	L.S.H.	.....	$\sim10^{-3}$	Softens at 50°C; use below 40°C; insoluble in usual organic liquids and acids
Dennison's sealing wax	J.G.B.	.....	$\sim10^{-5}$	Softens 60–80°C; soluble in alcohol
Lubriseal	A.H.T.	40	$<10^{-5}$	For general use
Lubriseal H.V.	A.H.T.	50	$\sim3 \times 10^{-6}$	
Myvawax S	C.V.C.	72.5	1×10^{-6}	Soluble in petroleum ether, CCl_4· C_6H_6
Myvacene S	C.V.C.	215	1×10^{-6}	Soluble in warm decalene
Picein	L.S.H.	.....	10^{-8}	Inert to usual organic liquids and inorganic acids; softens at 50°C
Vacuseal, light	Cenco	50	10^{-5}	
Vacuseal, heavy	Cenco	60	10^{-5}	
Cello-seal	E.A.	.....	10^{-6}	Soluble in $CHCl_3$
Cello-grease	E.A.	120	$<10^{-6}$	Soluble in $CHCl_3$

J.G.B.: James G. Biddle, Philadelphia.
C.V.C.: Consolidated Vacuum Corp. (now Bendix Corp.), Rochester, N.Y.
L.S.H.: Laboratory Supply Houses.
A.H.T.: Arthur H. Thomas Co., Philadelphia.
Cenco: Central Scientific Co., Chicago.
E.A.: Fisher Scientific Co., St. Louis.

TABLE 3.2 Characteristics of Gasket, Insulating, and Miscellaneous Materials

Trade or chemical name	Supplier	Vapor pressure at 20°C, torr	General application notes
Natural rubber	....	$\sim 10^{-5}$	Good elastic properties for gaskets; soluble in various common solvents, mineral oil
Butyl rubber	....	$<10^{-6}$	Gasket material; good resistance to common solvents and mineral oil; low permeability rate
Silicone rubber SE450	G.E.	$\sim 10^{-5}$	Useful as gasket material at elevated temperatures (up to about 400°F)
Neoprene rubber	B.F.G.	$\sim 10^{-5}$	Gasket material; good resistance to common solvents and mineral oil
Hycar rubber	B.F.G.	$<10^{-5}$	Gasket material; good resistance to common solvents and mineral oil
Myvaseal rubber	C.V.C.	$<10^{-5}$	Gasket material; good resistance to common solvents
Lucite	P.C.	$<10^{-5}$	Thermoplastic; very good working properties; useful optical properties
Nylon	E.I.D.	$\sim 10^{-5}$	Electric insulator
Teflon	E.I.D.	$<10^{-6}$	Chemically inert; used as insulator and for gaskets; cold-flows at room temperature; can be used above 500°F
Pyralin	E.I.D.	Large	Electric insulator; not advisable to use inside vacuum systems; thermoplastic
Polythene	E.I.D.	$\sim 10^{-6}$	Electric insulator; high dc resistivity
Polyethylene	P.C.	$\sim 10^{-6}$	Electric insulator; high dc resistivity
Polystyrene	P.C.	$\sim 10^{-6}$	Electric insulator; thermoplastic; very high dc resistivity
Kel-F	M.W.K.	$<10^{-6}$	Electric insulator and gasket material; chemically inert
Cellulose acetate	P.C.	$\sim 5 \times 10^{-6}$	Thermoplastic; good working properties
Micarta	W.E.	$\sim 10^{-4}$	Thermosetting; electrical insulating material
Silastic	D.C.	$\sim 10^{-5}$	High tensile strength; high-elongation silicone rubber stock
Garlock No. 8773	G.P.	$\sim 10^{-5}$	80 Durometer silicone rubber gasket material; suitable for use at elevated temperatures
Mycalex	M.C.	$\sim 10^{-6}$	Electrical insulating material
Formica XXBP	F.I.C.	Large	Electrical insulating material; generally not suitable for use in vacuum systems
Formica LE	F.I.C.	Large	Electrical insulating material; generally not suitable for use in vacuum systems
Glyptal, red GE 1201 (air-dried 5 days)	G.E.C.	$<5 \times 10^{-6}$	Red enamel; for sealing pipe-threaded and other types of joints; sometimes used to seal leaks temporarily
Viton	L.S.H.	10^{-9}	Gasket material

G. E.: General Electric Co., Waterford, N.Y.

B.F.G.: B. F. Goodrich Co., Akron, Ohio.

C.V.C.: Consolidated Vacuum Corp., Rochester, N.Y. (Now Bendix Corp.)

P. C.: Plax Corporation, West Hartford, Conn.

E.I.D.: E. I. du Pont de Nemours & Co., Wilmington, Del.

M.W.K.: M. W. Kellogg Co., Jersey City, N.J.

W. E.: Micarta Division, Westinghouse Elect. and Mfg. Co., Trafford, Pa.

G.P.: Garlock Packing Co., Palmyra, N.Y.

M.C.: Mycalex Corp. of America, Clifton, N.J.

F.I.C.: Formica Insulation Co., Cincinnati, Ohio.

G.E.C.: Chemical Division, General Electric Co., Pittsfield, Mass.

L. S. H.: Laboratory Supply Houses.

TABLE 3.3 Characteristics of Some Vapor-pump Fluids

Fluid	Vapor pressure at 25°C, torr	Molecular weight	Viscosity at 80°F, centipoises
Silicone DC-704*	2×10^{-8}	484.0	40.0
Silicone DC-705*	3×10^{-10}	546.0	186.0
Convaclor-12†	2×10^{-4}	326.0	4,520.0
Convalux-10†	2×10^{-9}	454.0 (avg)	1,000.0
Convoil-20†	8×10^{-6}	400.0	80.3
Octoil-S†	5×10^{-8}	426.7	18.2
Octoil†	2×10^{-7}	390.5	51.5
Water	20	18.0	
Mercury	2.5×10^{-3}	200.6	

* By permission from Dow Corning Corporation, Midland, Michigan.

† By permission from Consolidated Vacuum Corporation (now Bendix Corporation), Rochester, New York.

has resulted in a great difficulty with numerous vacuum components which have been wired inside high-vacuum systems with standard insulated wire. The difficulty is not just that the insulations used contribute substantial amounts of gas to the system and hence make it more difficult to achieve low pressures, but that the insulation gradually degrades, eventually leading to electrical shorts.

Of the available plastics, the best is Viton-A (DuPont). This material is useful down to the middle of the 1×10^{-10} torr range; but if used in the as-received condition, it contributes a considerable amount of initial outgassing. This can be largely overcome by a preliminary bakeout at approximately 250°F in a vacuum chamber which removes the unreacted plasticizers, catalysts, and water vapor from the Viton-A, rendering it more useful in ultrahigh-vacuum systems.

Vapor-pump fluids for use in vacuum systems must be tailored closely to the type of pump in which they are to be used. The basic properties of the currently available fluids are given in Table 3.3.

3.4 *Vapor Pressure of Metals*

We are accustomed to thinking of metals as having negligible vapor pressures, which is only approximately true. In many vacuum systems, considerable heat is produced, as in vacuum furnaces; and under these conditions, even common metals do become critical since they tend to volatilize at temperatures below the melting points. Figures 3.1 and

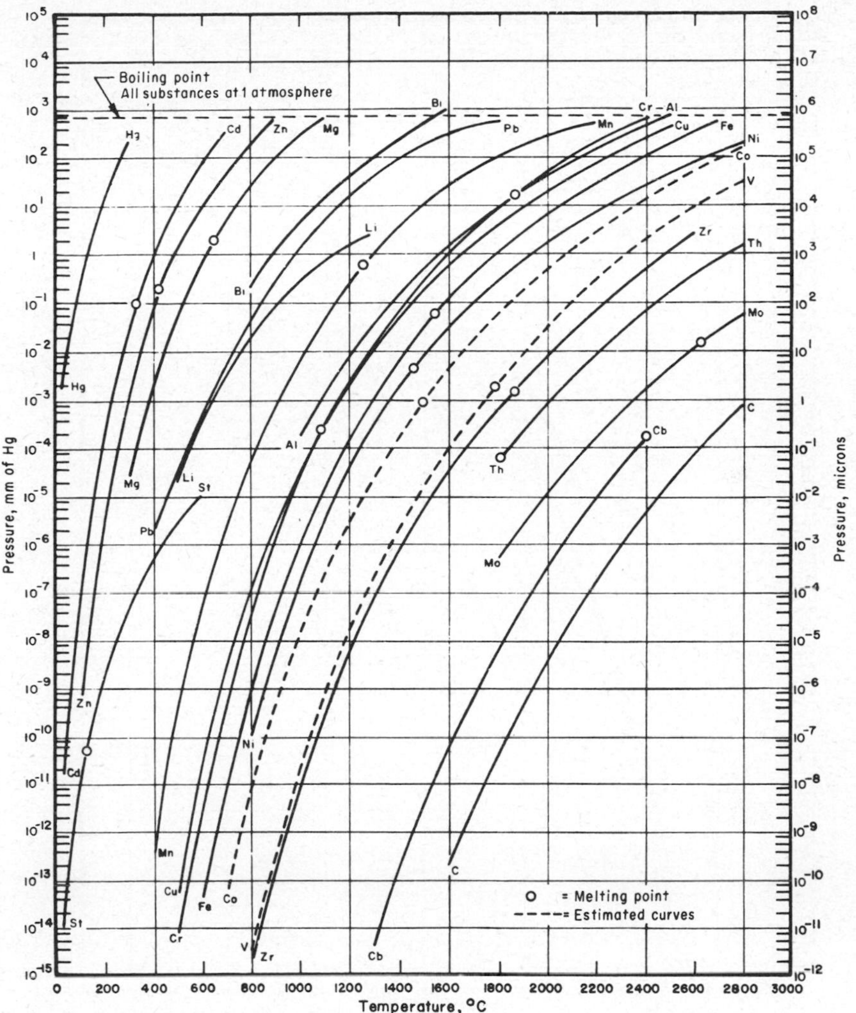

Fig. 3.1 Vapor pressure versus temperature. (*By permission of William E. Winter, Compilation of Data of Interest in Vacuum Metallurgy, Research Rept. 60-8-01-02-R4, Westinghouse Research Laboratories, Westinghouse Electric Corporation, East Pittsburgh, Pa., April 22, 1955.*)

3.2 show vapor pressure versus temperature curves at equilibrium for a number of common metals. It will be noted that at temperatures as low as 400°C, which is reached during bakeout of many systems, appreciable vapor pressures would render materials such as mercury, cadmium, zinc, magnesium, and lithium marginal in operation or totally impractical. It is for this reason that brass, an alloy of copper and zinc, cannot normally be used in ultrahigh-vacuum systems. In such

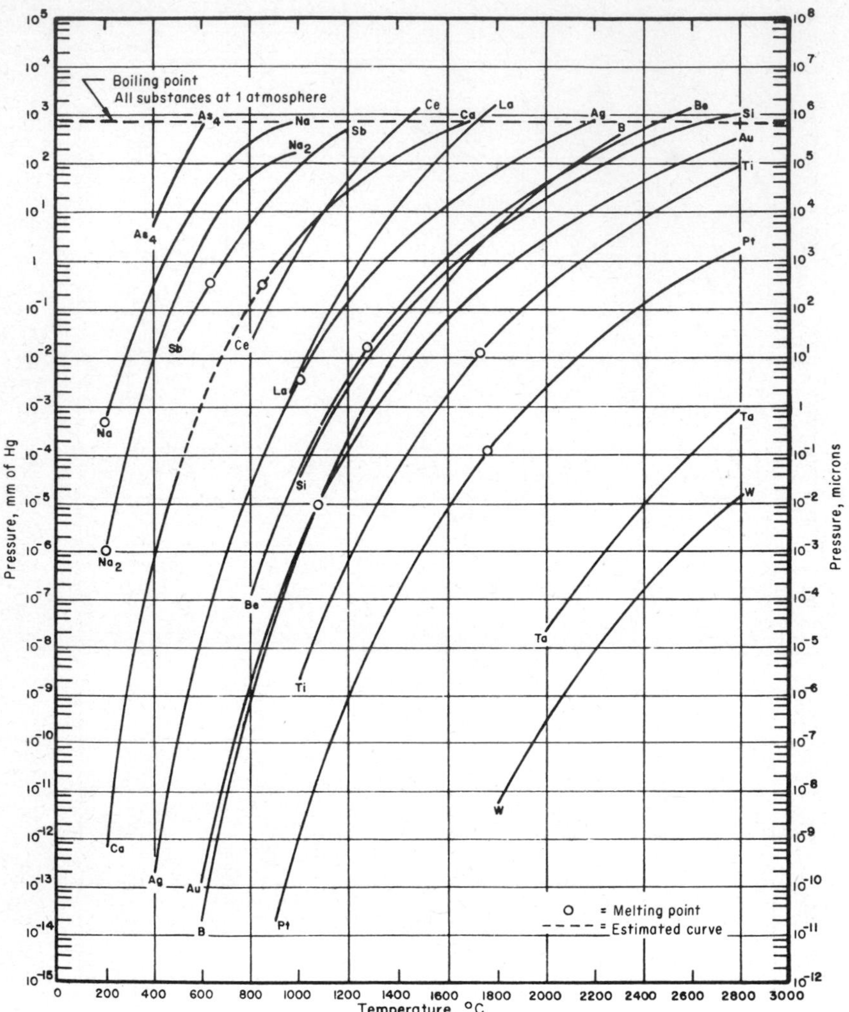

Fig. 3.2 Vapor pressure versus temperature. (*By permission of William E. Winter, Compilation of Data of Interest in Vacuum Metallurgy, Research Rept. 60-8-01-02-R4, Westinghouse Research Laboratories, Westinghouse Electric Corporation, East Pittsburgh, Pa., April 22, 1955.*)

a case, the difficulty arises from volatilization of zinc contained in the alloy which at slightly elevated temperatures begins to evaporate from the surfaces. From the curves, it is easy to see why materials such as molybdenum, columbium, carbon, tantalum, and tungsten are preferred for interior parts on high-temperature vacuum furnaces operating at temperatures of 2000 to 3000°C (3632 to 5432°F). Cadmium, often used on screws and other parts for rust protection, is especially bad in vacuum because of its high vapor pressure.

Where metals occur as alloys or compounds, they do not behave precisely in accordance with the curves shown in Figs. 3.1 and 3.2, which are for pure metals at equilibrium with saturated vapor in the surrounding atmosphere. However, the tendency in such compounds is for the more volatile elements to vaporize out and deposit elsewhere in the system. A dynamic system in which the pump operates continuously also does not approach the condition of equilibrium, and volatilization can continue unchecked as long as the volatilizing element is present. Thus in systems fabricated of nickel-chromium alloys (such as Inconel and stainless steel) used where temperatures are above 1800°F, we frequently observe the plating out of pure chromium on cooler parts of the system. This generally does no harm if it occurs on metallic parts, but can cause electrical shorting when the plating out occurs on electrical insulators. Manganese also plates out from normal low-alloy steels when heated to moderate temperatures because of its relatively high vapor pressure.

Vacuum furnaces are sometimes used for brazing operations using moderately high temperatures in vacuum, where the brazing alloys being melted contain high-vapor-pressure constituents. Under these conditions, some of the high-vapor-pressure constituents of the brazing alloy will volatilize out of the brazing metal and, in extreme cases, may prevent proper brazing through impoverishment of the joining material. Copper and manganese, used in many of the brazing alloys, are particularly bad in this effect, as is lithium, often used as a cleanup agent. For these reasons, the pressure must be controlled within brazing furnaces at a relatively high value, carefully adjusted to match the vapor pressure of materials which might volatilize from the brazing alloy. If cadmium occurs in the brazing alloys, it causes severe volatilization problems.

3.5 *The Effect of Vacuum on Metallic Oxides*

In ordinary brazing operations, it is conventional to make use of hydrogen as a reducing agent in order to reduce the metallic oxides always present on even the most carefully cleaned parts before the actual melting of the brazing materials. In some cases, a cleanup agent such as lithium is used to assist in this process of removing oxides from the surface of mating materials.

In high-vacuum systems, many of the common oxides which are present break down spontaneously without the use of any reducing agent. Figures 3.3 and 3.4 show the partial pressure of oxygen versus temperature for a number of materials. It will be noted that, where some of these oxides (principally those of copper and iron) break down relatively readily, certain others (particularly those of calcium, magnesium, aluminum, and silicon) will not break down at any reasonable combination of pressure and temperature.

The use of vacuum as an atmosphere in any high-temperature furnace not only prevents oxidation but also degasses the materials being

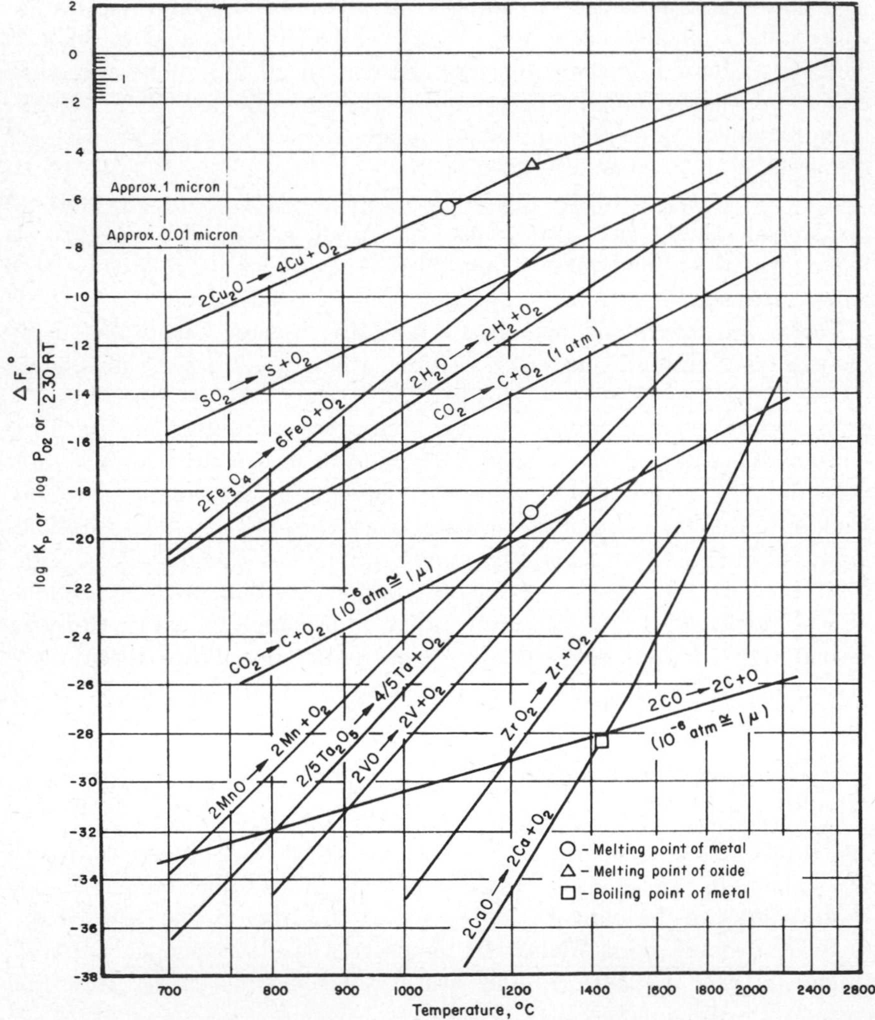

Fig. 3.3 Free energy of dissociation of metal oxides. (*By permission of William E. Winter, Compilation of Data of Interest in Vacuum Metallurgy, Research Rept. 60-8-01-02-R4, Westinghouse Research Laboratories, Westinghouse Electric Corporation, East Pittsburgh, Pa., April 22, 1955.*)

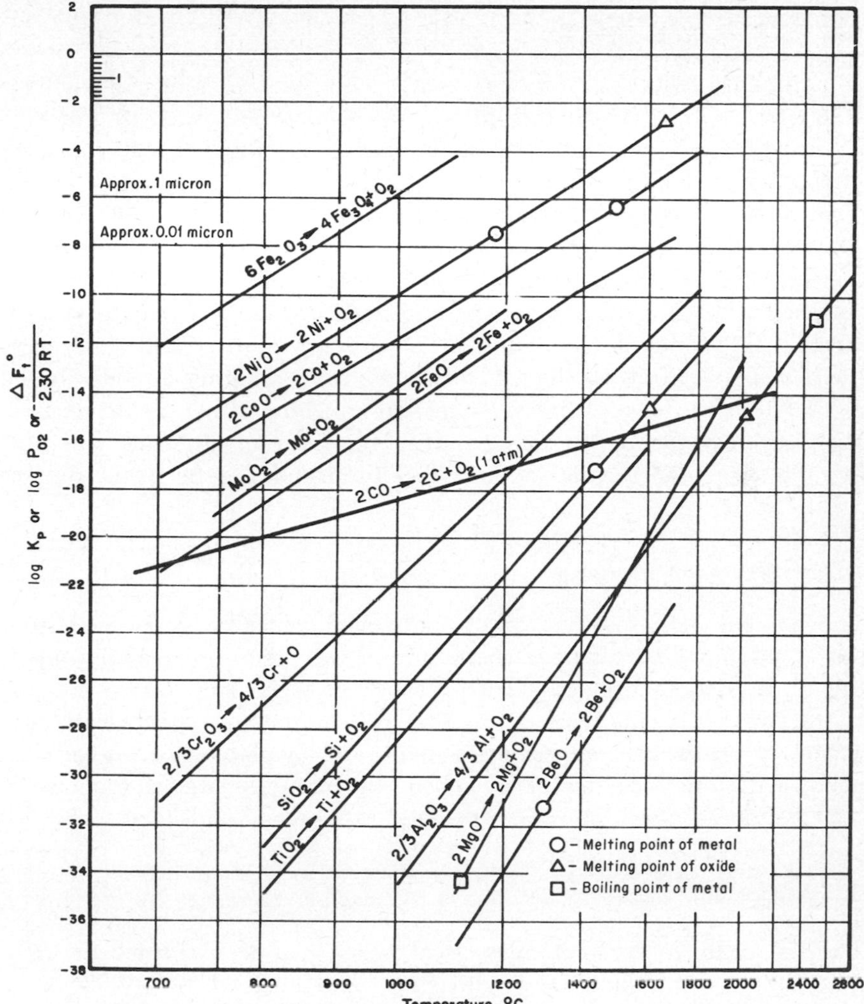

Fig. 3.4 Free energy of dissociation of metal oxides. (*By permission of William E. Winter, Compilation of Data of Interest in Vacuum Metallurgy, Research Rept. 60-8-01-02-R4, Westinghouse Research Laboratories, Westinghouse Electric Corporation, East Pittsburgh, Pa., April 22, 1955.*)

treated. In some cases, intergranular hydrogen or oxygen or high-vapor-pressure intergranular compounds can be removed, thus improving the physical properties of the materials being treated.

In brazing, the vacuum atmosphere frequently improves the flowability of the brazing material, thus improving the penetration of the material into the joints, while simultaneously requiring better stopoff materials to prevent flow where it is not desired.

3.6 *Adsorption Effects*

In addition to the basic vapor pressure of the materials of construction, all materials have characteristic effects in adsorbing surface layers of water vapor, oxygen, nitrogen, and other gases from the surrounding atmosphere. The presence of even small amounts of dirt and grease from fingerprints or other contamination on the surfaces can increase the amount of water vapor adsorbed by several orders of magnitude.

When the system is pumped down, these adsorbed gases from the surfaces of the vessel and the parts placed within it begin to desorb, initially rather rapidly but later at a very slow rate. This desorption phenomenon can go on almost indefinitely, representing at the extreme a limitation to the ultimate vacuum that can be attained in the system. Even after weeks of pumping, outgassing from metal walls may still be quite large and present serious difficulty in pumping down the system.

3.7 *Effects on Metals*

The amount of gases that will be adsorbed and slowly released from metal surfaces depends upon the nature of the metal involved and upon the surface of that metal—that is, whether it is oxidized or not. It is extremely difficult to measure the true outgassing effects with any accuracy. However, approximate data are available for these phenomena. Typical values after 1 hour of pumping are as follows:

Rusty steel......................	4.08 micron-liters/(sec)(sq ft)
Mild steel (clean)................	0.50 micron-liters/(sec)(sq ft)
Stainless steel (clean)............	0.16 micron-liters/(sec)(sq ft)

The last value refers to stainless steel which has been cleaned by pickling, followed by shot- or sandblasting, and cleaned with acetone or alcohol prior to beginning the pumping. The value can be reduced somewhat by grinding or polishing the interior surfaces so that the small deposits of oxide present on cleaned stainless steel can be reduced. However, despite the best preparation, values greatly lower than these are not achievable.

The extremely high values for rusty steel and the relatively high values even for clean steel indicate why high-vacuum systems and ultrahigh-vacuum systems cannot be effectively fabricated from ordinary steel materials. Every time such a system is opened to the atmosphere, some of the moisture present in the air results in the creation of a small amount of rust on the steel surfaces. This, in turn, holds down many times its own weight of water vapor which will be evolved when the system is reclosed and pumped down. The resulting prolongation of pumpdown time is so great that such vessels are almost universally fabricated of stainless steel materials.

The values given above for stainless steel represent the initial values which decay as pumpdown proceeds. For the first few hours of pumpdown, this outgassing rate appears to decay at room temperature linearly with time. However, it would appear that the decay rate eventually comes to equilibrium values such that pressures much below 1×10^{-8} torr cannot be attained without some form of auxiliary treatment.

To achieve further pumpdown, it therefore becomes necessary to resort to either heating or cooling the surfaces (as described in Sec. 3.9). When heat is applied, the evolution rate of the adherent gas molecules greatly increases, with the result that the pressure will normally rise as heating begins. A very effective treatment during this period of initial outgassing the chamber is to hold pressure at about 200 microns (2×10^{-1} torr) utilizing a mechanical pump, by the admission of a very small amount of dried nitrogen gas obtained from a liquid nitrogen dewar and admitted to the chamber at a point as distant as possible from the pumping port. This process is termed "gas bleed degassing." The flow of this gas within the chamber while the motion of the gas molecules is still viscous in nature assists in the removal of water vapor and other contaminants evolved from the chamber walls and causes the evolved material to flow into the pump and be discharged outside.

A practice that has turned out to be highly desirable for a new system before testing begins involves a bakeout period of approximately 8 hours at a temperature of 300 to 400°C with the gas bleed on, followed by a period of 16 hours with the gas bleed off, with the diffusion pump operating and the heat continued for the entire period. During this period, it will be observed that the pressure gradually lowers, generally reaching values of approximately 1×10^{-7} torr or lower by the end of the 24-hour bakeout period. If the pressure is plotted against time during the bakeout period, an appropriate point to terminate bakeout can be determined by the slope of the curve becoming zero (horizontal). Routine plotting of all bakeout pumpdown curves will often save considerable time in baking out systems.

At the end of the bakeout period, the chamber is allowed to return to room temperature while pumping continues. Normally, a pressure drop of 1 to 2 decades will occur as the temperature returns to room temperature. A further pumping period of perhaps 8 hours will usually take the chamber to approximately its ultimate pressure.

This lengthy bakeout period of 24 hours, as just described, is recommended for the initial preparation of a new chamber for service. However, after the chamber has once been preconditioned in this fashion, such lengthy bakeouts will not again be necessary unless the system has become badly contaminated. It would appear that the initial outgassing overcome by the long bakeout is primarily that due to surface effects on the inside of the vessel induced during the melting, rolling, and fabrication of the stainless steel sheet of which the vessel is made, and that these effects can only be removed by the process described. In addition, relatively large amounts of hydrogen seem to be contained interstitially in the metal; these gradually diffuse to the surface and are evolved. It is desirable to remove the hydrogen from at least the surface layer during bakeout, thus reducing the amount which will later evolve to that which can diffuse to the surface at room temperature. Moreover, subsequent bakeouts can be limited to periods of from 2 to 4 hours after the chamber has been opened and exposed to air, provided no extensive contamination has taken place. In fact, for some purposes where heat cannot be applied during actual testing, pressures rather closely approaching the ultimates attained with bakeout can be reached after somewhat longer pumpdown times without the use of heat at all.

3.8 *Plastics and Organic Materials*

None of the procedures outlined above can be employed for plastics and organics, since sufficiently high bakeout temperatures to remove gases would destroy the materials. For this reason, such materials must be used in the as-received form, but only after careful cleaning with such cleaning agents as are compatible with the materials. If possible, it is a good idea to heat the plastics to a slightly elevated temperature in a preconditioning chamber under vacuum in order to get the worst of the contamination off before they go into the ultrahigh-vacuum system. Other than this, the only solution lies in the choice of low-vapor-pressure materials and the brute-force technique of pumping until they reach the desired pressure.

3.9 *Degradation Effects*

Certain organic compounds, in the form of plastics or greases, are subject to long-term degradation effects. These may occur due to the volatilization of some element of the composition, such as plasticizers, thickeners, or catalysts. In other cases, the compound may actually break apart at very slightly elevated temperatures and at low pressures. The products so evolved not only add a gas load to the system but can contaminate the system in such a fashion as to adversely affect tests or operations being performed. In addition, the nature of the substance undergoing degradation can change radically. Thus lubricants can be turned into grinding compounds, and complex plastics can either become hard and brittle, resulting in cracking failure, or soften, turn gummy, and fail through loss of integrity.

These remarks are not meant to indicate that all plastic materials are useless in vacuum. Rather it is intended as a warning that vacuum and its effects must be considered when such materials are to be used.

In general, the outgassing rates for plastic materials vary from 1 to 2 micron-liters per second per square foot after 1 hour of pumping. There are, however, a few materials that have lower rates than these, ranging from 0.4 to 0.5 micron-liters per second per square foot after 1 hour of pumping.

The outgassing effects from plastics generally do not decay linearly with time as do the effects from metals, but usually decay approximately as the square root of the time, thus greatly prolonging the outgassing periods.

3.10 *Cryogenic Effects*

Where bakeout is not permissible due to certain of the characteristics of the materials, the outgassing effects mentioned can be inhibited by cooling the surfaces to low temperatures. In general, outgassing can be substantially eliminated from most materials if these may be cooled to temperatures of 100°K or below, as is the normal result of using liquid nitrogen. Under these conditions, the outgassing rate is so extremely low as to be difficult to measure, and no firm values are available. However, the technique has proved extremely valuable in many systems and is recommended where the function of the material is not affected by its low temperature. Unfortunately, this technique cannot, in general, be applied to wire insulations and plastics used in various space vehicles or experiments.

It should be noted, as is indicated in Chapter 8, that some of the gases condensed on 100°K cold surfaces at high vacuum pressures may reevaporate if and when pressures reach ultrahigh-vacuum level, which sometimes presents problems.

3.11 *Time Effects*

In the preceding discussion of bakeout, we used the optimum conditions of a temperature of from 300 to 400°C and a time of 24 hours. However, it sometimes happens that, because of the fragility of various portions of the vacuum system or of tests contained therein, temperatures cannot be raised to this value. This is particularly true where aluminum parts are used, which become so soft as to distort seriously at 400°C.

Even a very moderate bakeout temperature of 100 to 200°C for 2 to 4 hours will often reduce the total outgassing by a factor of 10, and can often be employed when higher temperatures are impractical.

Bear in mind also that parts may be, and often are, heated during experiments to temperatures higher than those employed in bakeout. In this case, additional outgassing will occur.

We shall give examples of the computational methods used for computing outgassing decay as pumping progresses in a later chapter of this book. Suffice it to say at present that, in those cases where no bakeout whatsoever may be used, it may be advisable to precondition the system while empty with as thorough a bakeout as possible and then to rely on time alone to secure low pressures after work pieces have been placed within the chamber. Times may be long if extremely low pressures are required. However, the decay rate will ordinarily allow the achievement of low pressures within a reasonable pumping time even under adverse conditions, provided the chamber itself has first been preconditioned by adequate bakeout.

3.12 *Alternate Methods of Degassing Chambers*

Since some experiments will not permit the use of high-temperature bakeout, alternate methods may be desired. Ion bombardment may be carried out at pressures somewhat higher than 1 micron (1×10^{-3} torr) if a high potential is applied to the chamber through inclusion of an insulated electrode connected to a high-voltage power supply. A 2,000-volt low-current power supply will yield a strong stream of ions under these conditions, with a vigorous glow discharge, which results in rapid removal of gas from the adsorbed condition on the surfaces, thus permitting its removal from the system. Unfortunately, at this pressure recontamination does occur, so that this method has not been used as a cleanup method for extremely low-pressure systems. It is, however, an extremely useful technique for moderately low-pressure systems, such as those used in vacuum coating, where the ion bombardment may readily remove contamination from even organic materials in a few minutes, permitting the operation to be carried out at pressures of approximately 1×10^{-4} or 1×10^{-5} torr.

chapter 4

Mechanical Vacuum Pumps

4.1 *Introduction*

Mechanical vacuum pumps are the only types (with the exception of steam ejectors) which will operate directly against atmospheric back-pressure and must therefore be used in roughing all vacuum systems except for those special cases where cryogenic sorption pumping is used. They must also be used as backing pumps to reduce the forepressure of diffusion pumps to the required value so that they can operate properly without backstreaming.

Special mechanically driven pumps of the Roots blower or booster types and of the molecular-drag types operate to lower pressures than the normally used mechanical pumps, but must themselves be backed by a mechanical pump as they are not capable of exhausting directly to the atmosphere.

4.2 *Vane-type Pumps*

One of the earliest and still most popular of mechanical pumps is the vane-type pump. Figure 4.1 illustrates the principle of operation.

Basically, the pump consists of an off-center cylinder rotating in a stationary cylinder, the rotor being provided with two or more spring-loaded vanes which are pushed outward against the wall of the stationary cylinder and which oscillate as the pump rotates. A spring-loaded poppet valve is provided to exhaust the gases to the atmosphere. This type of pump has the advantage that the clearance between the vanes and the cylinder wall is held automatically at an extremely low value due to the spring action pushing the end of the vane against the wall. However, the side clearances, which are initially set at a very close value, are not automatically adjustable and therefore constitute a leakage area which must be sealed by means of the pump fluid. As the pumps rotate, the vanes act to compress the gas coming in from the entrance port to a point at which it can be ejected through the exhaust valve.

Such pumps are made in both single-stage and two-stage designs. The single-stage design is capable of reaching blankoff pressures on the order of 10 microns (1×10^{-2} torr), but is generally useful to working pressures of not less than 20 or 25 microns (2 to 2.5×10^{-2} torr). The double-stage pumps are capable of reaching blankoff pressures of 1×10^{-4} torr, but are generally useful down to pressures not lower than 5×10^{-4} torr, even under most favorable conditions—and under average conditions, not below 1×10^{-3} torr.

Such pumps are made by Kinney, Sargent-Welch, Cenco, Leybold-Heraeus, and others, and are in widespread use for speeds not greater than 50 cubic feet per minute at present. Figure 4.2 indicates the pumping speeds of various pumps in the Sargent-Welch line. From this figure it can be noted that the speed drops markedly as the pump approaches

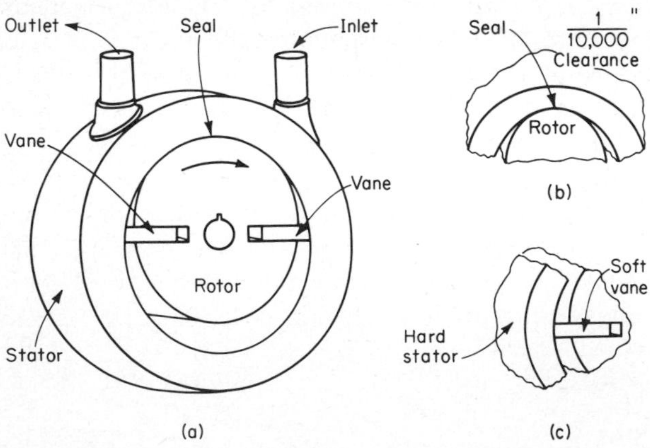

Fig. 4.1 Vane-type mechanical pump: (*a*) diagram of the movement; (*b*) view showing seal; (*c*) view showing vane construction. (*Sargent-Welch Scientific Co., Skokie, Ill.*)

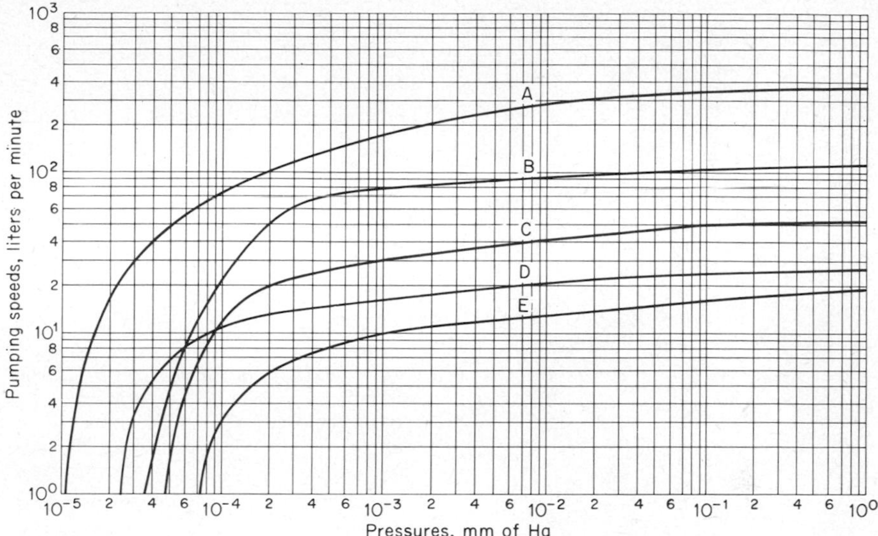

Fig. 4.2 Pumping-speed curves for various two-stage rotary vacuum pumps. (*Sargent-Welch Scientific Co., Skokie, Ill.*)

its ultimate limit. It is therefore very important, when selecting one of these pumps for a given job, to take note of the dropoff in pumping speed at the lower pressures and make due allowance for it. In general, where pressures below 1 micron (1×10^{-3} torr) are required, it is usually more economical to combine a smaller-size mechanical pump with some type of diffusion pump rather than attempt to use the mechanical pump alone, since the cooperating diffusion pump is much less expensive than an equivalent amount of capacity in the low-pressure range of the mechanical pump.

Such vane-type pumps have long useful lives and, in the smaller sizes, are almost universally used for roughing and backup pumps on vacuum systems. Their chief weakness lies in the close clearances involved, which can give trouble when contaminants enter from the system being pumped, requiring cleaning and reassembly of the pump. The reassembly operation on the vane-type pumps calls for a very considerable degree of precision, since any errors in setting the clearances will result in very serious degradation of pump performance; this, in turn, makes pump cleaning and repair difficult.

4.3 *Rotary-type Pumps*

The great bulk of vacuum pumps used on systems of medium and large size are of the rotary type, as built by Kinney, Stokes, Leybold-Heraeus,

Fig. 4.3 Rotary-piston-type vacuum pump: intake. (*Kinney Vacuum Co., Boston, Mass.*)

and many others. Basically, as shown in Figs. 4.3 and 4.4, they consist of a rotary eccentric revolving within a water-cooled or air-cooled circular housing connected to a reciprocating port which seals automatically during the compression stroke.

Pumps of this type are made for a wide variety of throughputs ranging from approximately 3 cubic feet per minute to 700 or more cubic feet per minute. They are made in both single-stage and double-stage designs, with pressure limits for the two types being about equivalent to those for the vane-type pumps described above.

These pumps, like the vane pumps mentioned earlier, are sealed by means of a lubricating fluid or oil which serves to seal the clearances between the stationary member and the rotating member, which does not touch either at the sides or radially. Because of the fashion in which oil is used as a sealing medium in these pumps, extremely close clearances are not required. Therefore, they are somewhat more tolerant of contamination than the vane-type pumps and are somewhat easier to clean, repair, and reassemble. The smaller-size pumps are air-cooled, and the larger sizes water-cooled because of the large amount of heat developed by the compression operation of the pumps.

In both types of pumps, the oil used as a sealing medium must serve also as a lubricant for the moving parts and therefore must be highly refined and compounded to perform these functions, while still retaining a very low vapor pressure.

The chief problems with both types of mechanical pumps usually arise due to contamination of the sealing oil by water vapor coming from the chamber being evacuated. Where water vapor is carried into the oil, it has the effect of raising the vapor pressure of the sealing medium, thus severely limiting the ultimate pressure the pump can reach and its throughput at all of the lower pressure ranges.

One of the ways of improving the performance of such pumps where water vapor is a problem is to make use of the ballast valve. This device is a small leak valve positioned so that it opens into the compression area after the gas therein has been sealed off from the entrance port but before it has been exhausted from the exhaust port. The effect is to raise the pressure in this area sufficiently high that the effective compression ratio in the pump is reduced to that which will permit the water to be ejected as a vapor along with the permanent gas. A mechanical pump roughing a large chamber filled with moisture-laden air may become so water-loaded as to raise its blankoff pressure to 100 microns (1×10^{-1} torr) or even 500 microns (5×10^{-1} torr) in extreme cases. Such a pump, of course, cannot back a diffusion pump until it has been freed of water or other high-vapor-pressure contaminants causing the trouble.

This can frequently be accomplished by blanking off the pump inlet and running it for 12 hours or so with the ballast valve open. A more rapid cleanup can be achieved if dry air (—40°F dewpoint) is admitted to the ballast valve instead of moist room air. An even more rapid cleanup can be achieved by the use of nitrogen gas from a cylinder of liquid nitrogen. Using this means, mechanical pumps using TCP

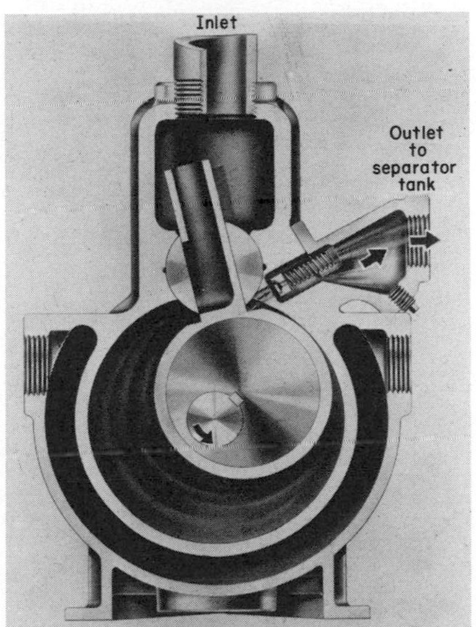

Fig. 4.4 Rotary-piston-type vacuum pump: exhaust. (*Kinney Vacuum Co., Boston, Mass.*)

(an oil used for oxygen compatibility, unfortunately highly hygroscopic) cleaned up from the original 500 to 600 microns (5 to 6×10^{-1} torr) to 40 microns (4×10^{-2} torr) blankoff in under 2 hours.

If it is known in advance that a considerable quantity of water vapor must be handled by the pump, the ballast valve may be opened a very small amount during the actual exhaust operation. This prevents the accumulation of water vapor within the system but imposes a higher ultimate pressure on the pump operation, so that one must be satisfied with a poorer performance where water vapor is involved. Some makes of pumps are provided with a small ballast orifice which is always open and which provides a small amount of ballast air under all conditions. An even better solution, of course, is to employ a liquid-nitrogen-cooled trap in the roughing train. This not only takes care of the water-vapor problem during roughing, but also greatly increases the pumping speed, especially during the low-pressure end of the pumpdown cycle.

Where minimum pressures are required, as indicated earlier, a double-stage pump is desirable. However, where considerable quantities of water vapor must be handled, a single-stage pump with the ballast valve slightly opened will handle this large amount of water vapor more effectively. It may therefore be indicated in some kinds of operations, in spite of the pressure limitations imposed thereby.

4.4 *Mechanical Booster Pumps*

It is sometimes desirable to make use of a pump capable of reaching lower pressures than the straight single- or double-stage mechanical pumps described above and having very high throughput at these low pressures. In addition, it may be desirable to use such a pump to prevent the migration of oil molecules from the mechanical pump into the system.

For these purposes, a Roots-type blower or booster is frequently employed. As shown in Fig. 4.5, this pump consists of two kidney-shaped eccentrics driven by a common drive and phased so that they interlock with one another to trap a compressible volume of gas and expel this at a higher pressure through the outlet. The lobes never touch each other or the casing and can be run without lubricant except for their bearings, which are isolated by seals. They therefore do not contribute any oil backstreaming effect, which may be important for some purposes.

Most such pumps are not capable of exhausting directly to the atmosphere, but must be backed by mechanical pumps of a conventional type. When staged with a mechanical pump of appropriate design, they can reach pressures at blankoff below 1×10^{-4} torr and can operate at high pumping speeds at pressures of the order of 5×10^{-4} torr. By

Fig. 4.5 Two stages of Roots blowers connected in series for very high-throughput pumping systems. (*Leybold-Heraeus, Inc., Monroeville, Pa.*)

staging two such blowers in series with a mechanical pump, even lower pressures can be reached, although their general field of usefulness lies in the range of 10 microns (1×10^{-2} torr) to 5×10^{-4} torr.

4.5 *Molecular-drag or Turbine Pumps*

There has recently been introduced to this country a new variety of mechanical pump, made by the Heraeus Company in Germany and marketed in the United States by Sargent-Welch, which is capable of achieving pressure as low as 1×10^{-9} torr. This pump makes use of a multiple-stage high-speed rotor consisting of a series of flat disks rotating between fixed disks which accelerate gas molecules by a collisional method to high speeds and utilize the centrifugal effect to compress these gas molecules. Such pumps operate at very high speeds of rotation (approximately 24,000 rpm) and have pumping speeds of 140 liters per second at their flanges. They are extremely expensive but form useful adjuncts for some types of laboratory equipment, especially roughing ion-pumped systems, because of their low-pressure capabilities. They must be backed by mechanical pumps of the two-stage variety. Figure 4.6 shows such a pump.

A larger-size drag pump having a speed of 260 liters per second has been available in Germany for some time and is now being marketed in the United States, thus enabling this useful type of pump to be used on systems requiring pumping speeds in excess of 140 liters per second.

Such pumps are inherently free of oil contamination effects, and actually act as most efficient traps for oil molecules backstreaming from mechanical pumps. This action results from the centrifugal effect, which pumps the heavy oil molecules much more effectively than the lighter permanent gases.

Where the pumping speed requirements are sufficiently low, such pumps (suitably backed by a mechanical pump) may be used for systems operating in the 1×10^{-7} to 5×10^{-8} torr range. Such a pumping system provides a very low-contamination system for mass spectrometers and various special laboratory systems where no contamination can be tolerated.

Fig. 4.6 Turbomolecular pump and backing mechanical pump. (*Sargent-Welch Scientific Co., Skokie, Ill.*)

4.6 *Water-sealed Pumps*

There are many types of vacuum systems where the primary requirement is evaporating large amounts of water. Such uses as vacuum drying, freeze drying, deaeration of condensers in power plants, prechilling by removal of water vapor from lettuce, fruits, etc., using the latent heat of vaporization to quickly prechill large shipments of produce, and the separation of chemical products by distillation at or near room temperature, all require the handling continuously of large amounts of water.

An ordinary oil-sealed rotary vacuum pump will quickly become inoperable under these conditions. Using steam ejectors, as has long been done for this purpose, requires large amounts of steam and cooling water, not always available. A rotary-type pump has been developed for such needs in which the sealing medium is water. The vacuum capabilities of such a system range from 30 torr when the water temperature is 85°F to 11 torr when the water temperature is 55°F. These pressures may be sufficient for some purposes. Where they are not, the water-sealed pump may be combined with a Roots-type blower, which makes ultimate pressures of 1 torr to 250 microns (2.5×10^{-1} torr) possible.

Such systems are finding widespread use in industry, especially in the dehydrated-food industry, which is growing rapidly.

chapter 5

Vapor-type Pumps

5.1 Introduction

The great bulk of vacuum systems in use rely on some type of vapor pump to achieve the high vacuum. The principle used here is most simply explained by considering a jet of some condensible vapor directed from an orifice in such a direction that there is a large component of velocity away from the system being pumped. Collision of the molecules of vapor with gas molecules imparts to the latter a velocity away from the system being pumped and toward the outlet of the pump. The vapor itself is then condensed on a cold surface and returned to the boiler for reevaporation and redirection through the jets. The noncondensible gas, having been given a velocity away from the system being pumped, is further accelerated by additional jets, thus creating a compression of the gas molecules which are finally removed from the pump by means of a backing pump which eventually exhausts the gas to the atmosphere. The efficiency of such a device depends upon the homogeneity of the vapor jet, the density of the molecules in the vapor jet, and their velocity. When properly designed, a four-jet system of this variety can achieve compression ratios of more than 1,000,000:1

in the gas being pumped and will operate against any form of gas molecule being evacuated from the system. Because of their universal characteristics, such systems are very widely used in almost all commercial vacuum systems.

5.2 *Vapor-type Ejector or Booster Pumps*

Pumps of this type are designed for operation at pressures intermediately between those attained by a mechanical pump and those attained by the more conventional diffusion pumps. Their principle of operation involves the ejection of mercury or oil vapors from one or more jets to accelerate gas in the preferred direction away from the chamber being pumped; the gas is finally ejected to the atmosphere through a mechanical pump of conventional type. Such pumps operate with special fluids enabling them to pump effectively at very high forepressures, generally up to 500 microns (5×10^{-1} torr) or above. The use of such fluids involves the use of higher boiler temperatures and limits the ultimate vacuum that can be attained by such pumps, generally to values of approximately 5×10^{-4} torr. However, they have very high pumping speeds in the intermediate pressure ranges; and, since they can be cut into the system at relatively high pressures, they provide a rapid means of reducing system pressures to a point where conventional diffusion pumps may be employed, achieving faster pumpdown rates than could be attained by a reasonable-sized mechanical pump. In other words, they allow the use of a smaller mechanical pump than would otherwise be required. The price of a booster is frequently less than the difference in price between the small backing pump required with the booster and the much larger pump required without it, so that an overall cost saving may result by interposing a booster stage.

Where very large diffusion pumps are required for the system, it is usually necessary to make use of either mechanical boosters of the Roots types as described earlier or oil-type boosters in order to reduce the backing required for the large diffusion pumps to a reasonable value. Therefore, large systems are of the three-stage type, employing a mechanical backing pump, a booster pump of either the Roots type (mechanical) or the oil type, plus the multiple-stage diffusion pump.

5.3 *Diffusion Pumps*

The multiple-stage diffusion pump has long been the workhorse of vacuum systems, being used at all pressures from approximately 1 micron (1×10^{-3} torr) down to 1×10^{-9} torr and below. Figure 5.1 shows the construction of a typical pump of this variety.

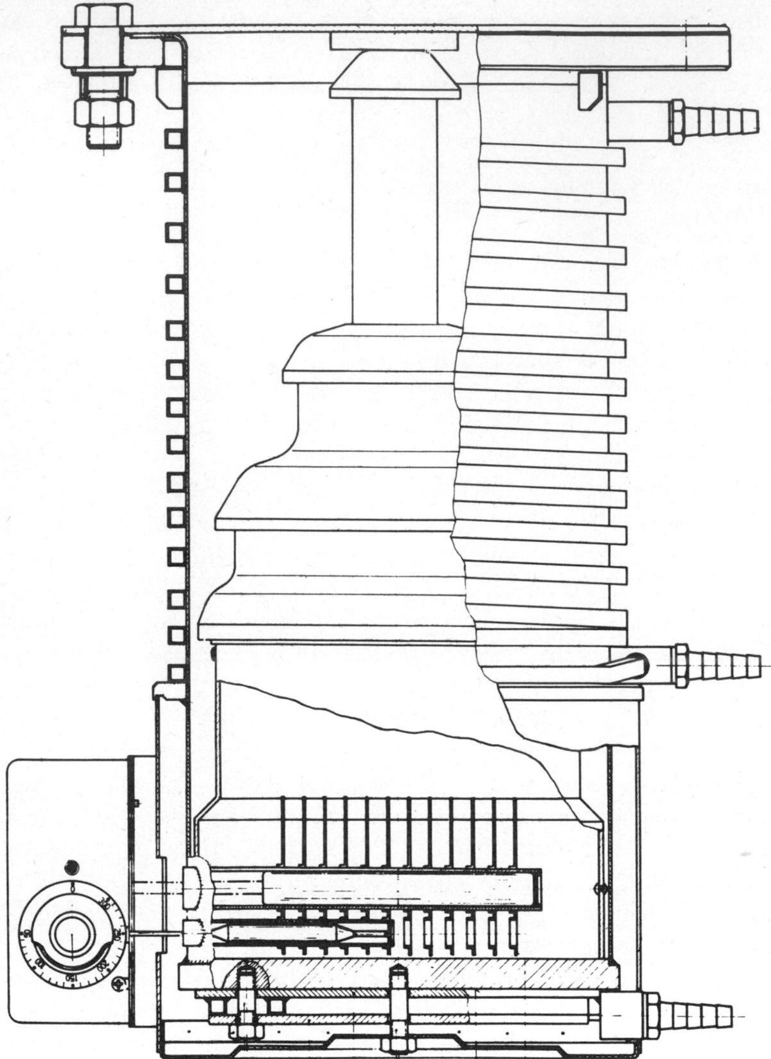

Fig. 5.1 Cross-sectional diagram of an oil diffusion pump. (*Leybold-Heraeus, Inc., Monroeville, Pa.*)

The ultimate pressure which can be reached by such pumps depends critically on the vapor pressure of the fluid being used, as measured at the pump entrance throat. The use of cold traps, either water-cooled or liquid-nitrogen-cooled, further reduces the vapor pressure to that equivalent to the cold-trap surface temperature and enables such pumps to reach lower ultimate vacuums.

All diffusion pumps suffer to a greater or lesser degree from the phenomenon of backstreaming—that is, the movement of oil molecules countercurrent to the gas being pumped, thus resulting in contamination of the system. Many studies have been made of the backstreaming phenomenon, and a number of causes thereof have been found. The first of these is the breakdown of the oil. All fluids used in pumps gradually deteriorate due to breakdown of some of the long chain molecules composing the fluid into fractions of lighter molecular weight. Some of these oil fractions have such low vapor pressures that they are not effectively trapped even by surfaces maintained at the temperature of liquid nitrogen (78°K), and therefore can migrate through the trap into the system. An effort to eliminate this phenomenon involves the use of the so-called "fractionating pump," in which the boiler area is divided into a number of concentric rings, the outermost ring being maintained at the lowest temperature and the innermost ring being maintained at the highest. Under these conditions, the oil molecules of the lowest boiling point will evaporate in the outermost ring and thence be fed into the lowest of the several jets of the pump, so that their opportunities for migrating upwards into the system are severely limited. Conversely, the highest-boiling molecules will be evaporated from the innermost ring, from which they will be carried to the uppermost jet, thus limiting the number of low-boiling-point molecules that enter the upper jet and that hence might migrate back into the system. In this fashion, the products of decomposition in the pump oil which have the lowest boiling points are confined essentially to the lowest jet, causing negligible backstreaming.

A second form of backstreaming results from the fact that the oil vapors entering the chimney are saturated and necessarily cool slightly in passing up the chimney to the jets. Because of this cooling effect, some of the oil vapor will condense into small droplets which tend to exit from the jets with much less velocity than the true vapor stream. These droplets can migrate upwards and get into the system or at least reach the cold traps. An effort to avoid this involves the use of heaters which project above the oil level, thus superheating the oil vapor slightly and preventing condensation until the jet has been formed and reaches the pump walls.

It has further been found that the use of a water-cooled cap just above the top jet, or, alternatively, a cap which cools by radiation to the liquid-nitrogen-cooled cold trap just above, will reduce the backstreaming by a large factor, since it tends to intercept those droplet particles which would otherwise pass upward out the inlet mouth of the pump.

A third form of backstreaming results from the fact that a molecule of the jet stream colliding with a gas molecule may acquire an upward velocity due to the collision, or may acquire such a velocity upon impinging on the molecularly rough chamber wall. In either case, a molecule

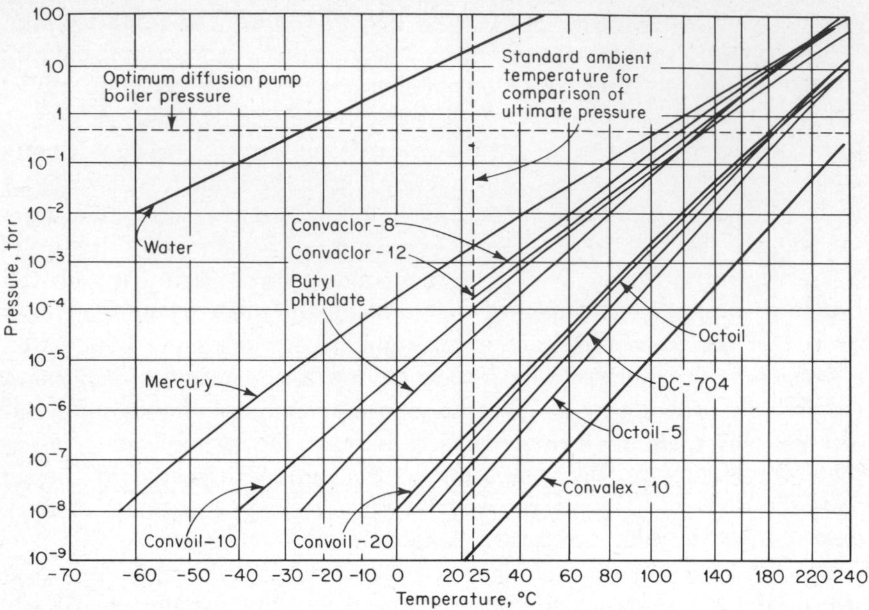

Fig. 5.2 Pressure-temperature characteristics of pump fluids. (*Bendix Corp., Rochester, N.Y.*)

of oil will acquire a velocity upstream toward the chamber and may reach this chamber if it is not stopped by a cold trap. A similar effect can occur within the cold trap itself, where oil-to-oil or oil-to-air collisions can enable some molecules to avoid contact with the cooled surfaces and thus pass through a "single-bounce" trap.

In general, much-improved trap designs have become available recently which, when providing a "double-bounce" geometry to migrating molecules of pump fluid, seem to stop effectively all backstreaming of pump fluids of the original molecular nature. The remaining backstreaming is then that due to noncondensible fractions of the pump fluid whose vapor pressure is such that they will not condense at the trap temperature and hence can migrate into the chamber. These traps also employ liquid-nitrogen-cooled walls to stop the creeping of oil molecules condensed on the trap walls which might otherwise reach the system.

The effects of the remaining backstreaming are small but can be serious for some types of usages, where even the very small amounts of hydrocarbons contributed to the chamber by the pump can cause undesirable effects within the working area. Mass-spectrometer readings of chambers pumped by very stable oils in diffusion pumps of good design will generally show hydrocarbons averaging approximately 2 percent of the total gaseous content of the chamber, unless the chamber is provided with internal cryogenic shrouds, in which case the readings on these molecules tend to reach even lower values. Figure 5.2 shows some of the fluids used in diffusion pumps and the temperature-pressure ranges in which they are useful (see also Table 3.3).

5.4 *Diffusion-pump Speeds*

The measurement of diffusion-pump speeds at high pressure levels is relatively straightforward, but becomes extremely difficult at low pressures. The method employed under a standard of the American Vacuum Society is that shown schematically in Fig. 5.3. A test dome is fabricated of stainless steel whose diameter is equal to the diameter of the pump throat and whose height is equal to 1½ times the diameter. The top is made sloping. The gas to be used in the calibration is admitted through a leak valve from a measuring device, and the jet is directed upward against the dome. A vacuum gauge is inserted into the test dome at a point just above the pump inlet. In the simplest form of this device, the metering system consists of a small graduated pipette

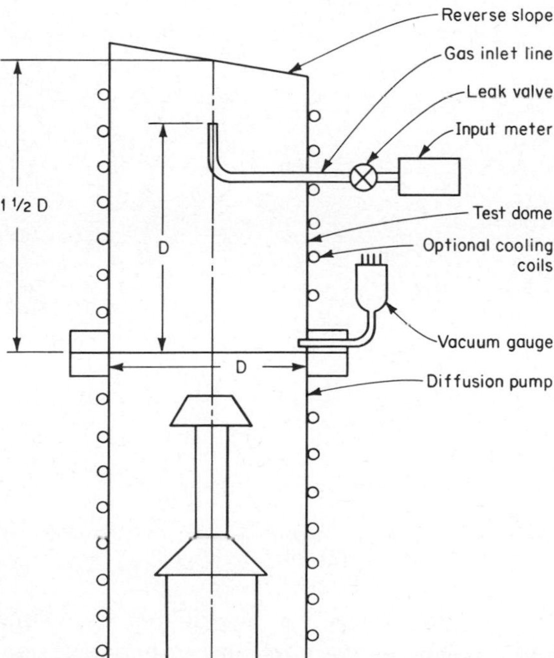

Fig. 5.3 Standard speed-measuring arrangement for diffusion pumps.

whose lower end is immersed in a large reservoir of pump fluid, preferably that being used in the pump.

The pump is started and allowed to run until equilibrium has been reached, admitting gas from the source (generally the atmosphere) through the small leak valve. When equilibrium has been established, the three-way valve is turned in such a fashion that the gas comes from the measuring pipette, being replaced by oil from the reservoir. In order to convert the measured volume of gas admitted from the reservoir into pump speed in liters per second, it is necessary to know the pressure in the speed dome; therefore, the calibration system is quite vulnerable to errors in the gauge reading. For accurate work, the gauge must be precalibrated by means of a standard gauge-calibration system before the pump speed can be accurately measured. Where the pressure is known, as well as the volume of the connecting piping, the following calculations can be used to determine the pumping speed:

$$v_1 = v_0 - L \tag{5.1}$$

$$p_1 = p_0 - h \tag{5.2}$$

which gives

$$\frac{p_0 v_0 - p_1 v_1}{t} = \frac{p_0 L + v_1 h}{t}$$

defining

$$Q = \frac{P_0 V_0 - P_1 V_1}{t} = P_2 S_p$$

then

$$S_p = \frac{P_0 L + v_1 h}{P_2 t} \tag{5.3}$$

where V_0 = initial volume of tubulation, valves and pipette
L = collected volume
P_0 = 760
h = static head (height of liquid)
P_2 = dome pressure
t = time

This technique is quite straightforward for the higher pressures. However, at the lower pressures, numerous errors make themselves apparent. The errors are of four principal types:

1. The true pumping speed of the pump will be equal to the amount

of gas admitted to the test dome plus the amount of gas being desorbed from the walls of the test dome. The desorbed volume is small in the range down to 1×10^{-5} torr, but becomes quite large at pressures of 1×10^{-6} and lower unless the test dome has been elaborately baked and preconditioned. Even under these conditions, some outgassing of the test dome remains, particularly at the lower pressures, thus invalidating speed measurements otherwise arrived at.

2. The contribution of gas due to the outgassing of the O rings at the bottom of the test dome and the permeation of gas through the O rings is being pumped by the diffusion pump in addition to the leak volume. This error increases greatly as the pressure is lowered; and unless very special pains are taken to eliminate this effect, the leakage values may overrun the amount of gas being admitted and measured. In general, at lower pressure levels refrigerated butyl rubber O rings would seem to be the only alternative to metal O rings heavily clamped. Even under the best conditions, however, some gas is contributed by leakage through the O rings or desorption from the O ring material, which is not metered and therefore causes errors in the readings.

3. At relatively high pressures, the amount of gas admitted from the pipette is relatively large, and thus the errors in reading the height of the oil in the pipette or errors in the calibration of same do not produce large percentage errors. At extremely low pressures, however, the multiplying factor is so large that the quantity of gas which is admitted becomes extremely small. This in turn calls for a pipette of almost capillary dimensions, which is difficult to calibrate accurately. The errors in reading the height of the oil are also multiplied in the small-diameter pipette, since capillary effects, sticking, etc., become large in proportion to the volume of gas measured. Moreover, the calibrated leak used for controlling the rate of admission of the gas becomes somewhat unstable at extremely low flows.

4. The gauge errors which enter directly into the calculations become quite large at low pressures. Means of calibrating gauges accurately are difficult for pressures below 1×10^{-5} or 1×10^{-6} torr. Any gauge error will enter directly into calculation of the speed of the pump, and hence produce correspondingly large errors in the throughput calculations.

Because of these various difficulties, pumps are not usually speed-tested at the lower ends of their possible ranges.

On theoretical grounds, it seems probable that the pump speeds indicated by the curves, which (as ordinarily stated) approach zero at pressures in the range of 1×10^{-8} or 1×10^{-9} torr, are in serious error. The apparent lack of pumping speed as measured at these pressures is simply the result of the accumulation of the types of errors specified above, which makes the speed-dome method of pump testing inapplicable to pressure ranges in which we are most interested.

From a theoretical design consideration, it can be shown that the speed of a well-designed diffusion pump should be constant at all pressures, being limited only by the diffusion rate of the gas molecules into the upper jet of the pump, which rate is independent of pressure at the lower pressure ranges.

That the pump speeds are indeed constant as pressures go down is apparent from work done in various places where it has been found that with extremely well-designed systems, thoroughly baked out, and with excellent construction of sealing O rings, passthroughs, etc., pressures in the range of 1×10^{-11} to 1×10^{-12} torr can in fact be obtained. This could hardly be true if the pump speeds approach zero in the fashion shown by the average pump-speed curve. Evidence indicates that diffusion-pump speeds are constant on a volumetric basis to pressures of at least as low as 1×10^{-12} torr. The failure to obtain such low pressures in the average system stems not from a lack of speed in the pump, but from leaks in the system, failure to outgas the system walls, leaks in seals and passthroughs, or failure of the gauge used to read very low pressures accurately.

Typical pump curves, as published in manufacturers' literature, are shown in Fig. 5.4, in which the sloping portion decreasing to zero at pressures on the order of 1×10^{-6} or 1×10^{-7} torr represents the speed measurements obtained as outlined above in an unbaked test dome without a trap. The horizontal portion represents the true speed of the pump which extends outward to very low pressure values.

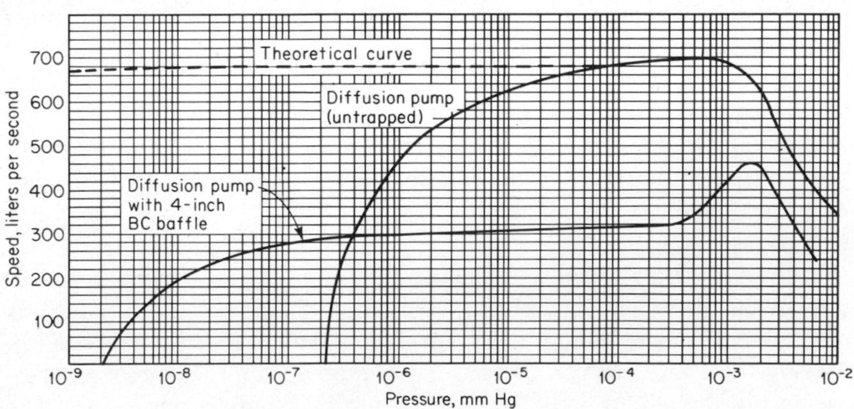

Fig. 5.4 Typical pump and baffle performance. (*Bendix Corp., Rochester, N.Y.*)

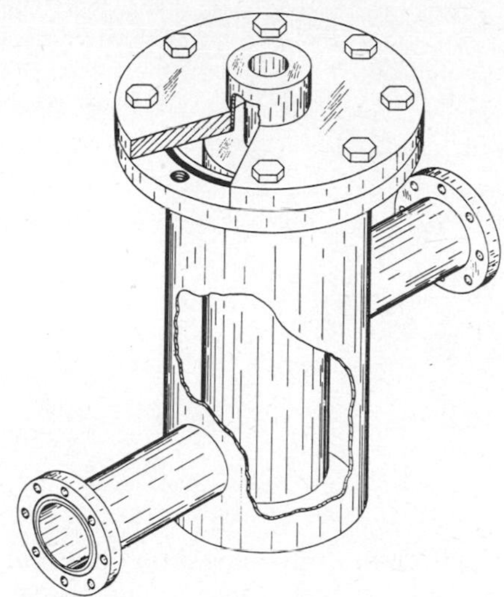

Fig. 5.5 Finger-type cold trap.

5.5 *Trapping Systems*

In order to achieve the extremely low pressures of which diffusion pumps are capable, it is necessary to limit the backstreaming from the pump to extremely low values. Three types of traps of varied sizes and efficiencies have come into use for this purpose. Figure 5.5 illustrates a common type of finger trap, in which a liquid-nitrogen-cooled cylinder extends into a larger section of pipe through which the gas flows between inlets and outlets which are not in line with each other. Such traps are normally used only for very small systems and are not extremely efficient, since quite an easy path exists by which a molecule of oil can travel through such traps without actually hitting the cold finger. Such traps are generally used, therefore, only between mechanical pumps and diffusion pumps and for other locations where viscous flow exists or where extremely high trapping efficiency is not required.

The chevron-type trap, shown in Fig. 5.6, is considerably more efficient in trapping oil molecules, especially if cold wall surfaces are provided to prevent migration of oil molecules along the wall by a creep mechanism. Such traps assure at least one contact of any migrating oil molecule with the chevrons, provided it does not experience an oil-to-air or oil-to-oil collision en route which would enable it to turn the corner through the chevrons without hitting the cold surfaces. Such traps provide reasonably good trapping efficiency plus fairly high conductance through the trap.

In computing the conductance of various trapping systems, it has turned out to be satisfactory to treat these as a straight length of pipe

Fig. 5.6 Chevron-type cold trap.

having a length equal to the actual length of the trap plus the additional length interposed by the labyrinth effect of the chevrons. The result, of course, is that in practical cases the conductance of the trap is chiefly determined by the actual flange-to-flange length of the trap and by its diameter.

As mentioned above, the plain chevron-type trap will trap all back-streaming oil molecules except those that experience an oil-to-air or oil-to-oil collision within the chevrons. However, the chances of such collisions are not zero, and some oil molecules will migrate through the chevrons and into the chamber. In an effort to overcome this effect, or at least reduce it by a large factor, the elbow-type trap, employing chevrons plus a cryogenically cooled elbow surface, as shown in Fig. 5.7, has been developed. Such traps have a trapping efficiency approximately 100 times better than a simple chevron trap, but suffer from the addi-

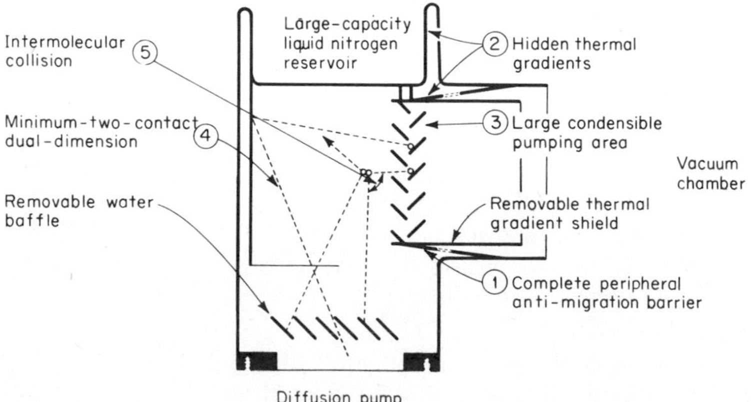

Fig. 5.7 Triple-bounce elbow trap. (*Aerovac Corp., Troy, N.Y.*)

tional length which results in a lower conductance. This effect can be minimized by increasing the diameter of the elbow sections to a value greater than that of the pump throat, thus regaining some of the conductance lost to the greater path length. Such traps will trap practically all the oil molecules that may enter their mouths, leaving only the noncondensible oil fractions to contaminate the system; these, when a suitably stable oil is used, represent a very small amount of contamination.

In practice, it is usually desirable to make use of a water-cooled baffle directly above the pump mouth which will collect the bulk of the oil molecules and allow them to drain back into the pump, thus avoiding depletion of the pump's oil supply. Following this water-cooled baffle, liquid-nitrogen-cooled baffles are essential if extremely clean systems are required, even when the most stable and lowest-pressure pump fluids are used. Pump fluids such as DC-704, DC-705, and Convalux-10 have lower vapor pressures than some of the hydrocarbon fluids previously used and will achieve lower pressures with poorer trapping. However, backstreaming still remains a problem with these fluids when they are employed in ultrahigh-vacuum systems.

It should be obvious from the above considerations that the net pumping speed available in the chamber is not a direct function of the speed of the pump as measured at its inlet flange. It is most important that the connecting piping between the pump and the chamber, including the length of the cold trap and any valves that may be used, should be kept to a minimum, since the impedances of these various elements severely limit the actual pumping speed attainable in the chamber. This in turn will limit the ultimate pressure which can be attained in the chamber. There have been many cases where the design of a pumping system has neglected this essential consideration and consequently performance of the system has been grossly limited because of the considerable distance interposed between the pumping flange and the chamber due to a wasteful design of interconnecting elements.

Unfortunately, diffusion pumps must be placed vertically in order to operate properly; therefore, it often becomes necessary to make use of an elbow to connect the diffusion pump to the chamber, at least where the height of the pump is such that the chamber would be unduly elevated above the floor level were the pump to be placed directly below without the elbow. Where such considerations exist, it is essential that the elbow employed be as short as possible and that, as far as possible, all trapping devices employed be placed within the elbow, thus avoiding additional lengthening of the path. It should be pointed out that in the molecular-flow regime nothing is gained by utilizing sweep elbows which are commonly employed in the viscous range. Instead, right-angle turns may be employed without introducing any additional impedance effects. It should also be borne in mind that it is often desirable to make the connecting piping larger in size than the pump flange itself, since the reduction in impedance thus achieved frequently well repays the extra cost in material for the larger elbow.

5.6 *Backing-pump Requirements*

In selecting a backing pump for a diffusion pump, consideration must be given not only to the required backing as stated by the diffusion-pump manufacturer but also to the effect of the size of the backing pump on throughput at the higher pressure ranges. In general, at low pressures the backing pump necessary may be very small indeed. In fact, in leak-checking systems, pumps as large as 48 inches in size have been backed by the leak detector alone when the system is at very low pressures.

However, at the higher pressures at which the diffusion pump operates, the throughput of the pump will be determined in large measure by the capacity of the backing pump, which must be relatively large in order to handle the large throughputs at pressures of 1 micron (1×10^{-3} torr) and above.

The failure to provide adequate backing under these conditions not only limits the throughput of the diffusion pump, and therefore the pumpdown time of the system, but also has the effect of increasing backstreaming, since an insufficiently large backing pump will cause breakdown of the upper jet stream, resulting in a large amount of oil backstreaming at the high-pressure end of the pumping curve. This effect usually takes place at the moment the diffusion pump is cut into the system, unless a very large backing capacity is available. Where backstreaming is critical, it is desirable to have sufficient capacity so that the pump passes through this critical backstreaming range very rapidly and achieves satisfactory jet stability within a minute or less of the time that it is cut into the line, thus limiting the actual volume of oil backstreamed to a tolerable level. This effect can, of course, be minimized or even completely eliminated on three-stage systems, since cutover of the diffusion pump can be held off until the backing ejector or diffusion pump has reached values of 0.5 micron (5×10^{-4} torr) or below, thus allowing the diffusion pump to come smoothly on stream without trouble.

It should also be pointed out that in systems where no valve can be used between the diffusion pump and the chamber and where, consequently, the diffusion pump must be heated up after roughing is complete, there will exist a period as the jets begin to operate when the jets will not have established a dense stream of vapor. At this moment, severe backstreaming occurs, and will continue until the boiler pressure has reached the design value and the jets are fully established. This, therefore, becomes a critical period for backstreaming on valveless systems. The same effect takes place when the system is shut down and once more the jet passes through the instability region as the boiler cools.

Some gain may be achieved in such a valveless system by holding flow in the viscous range during the heating of the diffusion pump. This may be accomplished by using a leak of dry air or nitrogen to hold the pressure in the micron (1×10^{-3} torr) range until the heaters of the diffusion pump have reached the normal operating temperature. When repressurizing the system, the diffusion pump continues to be heated while the leak brings the pressure into the viscous range, then is allowed to cool while the backing mechanical pumps continue to operate. After a short cooling period, the system may be shut down completely and repressurized. The purpose of this method is to sweep the oil vapor away from the chamber with a relatively large, viscous flow of purging gas.

One way of avoiding the contamination problem due to the fractionation of organic pump fluids is to make use of mercury. This fluid, being an element and not a compound, cannot break down. However, it has an inherently high vapor pressure and can only be well trapped by a double trapping system, one of whose traps must be liquid-nitrogen-cooled. Furthermore, mercury-type diffusion pumps are commercially available only in rather small sizes. Glass mercury pumps, however, are widely used in laboratory work.

By employing ejector or booster stages either placed between the mechanical pump and the diffusion pump or built into the diffusion pump, higher mechanical-pump backing pressures can be permitted without danger of excessive backstreaming. Certain modern designs of diffusion pumps have an ejector stage built into the pump sidearm, thus reducing the backing-pump requirements and the backstreaming problem.

chapter 6

Ionic and Sublimation Pumps

6.1 *Introduction*

It had long been known that in cold-cathode gauges of the Penning type gas pumping occurred due to sputtering of metallic particles from the plates. It remained for Hall, Jepson, and their collaborators to turn this phenomenon into a practical commercial pumping device. In ion pumps, as made commercially by Varian, Ultek, General Electric, Hughes, and others, an electronic discharge is set up between anode and cathode plates of which at least the cathode is fabricated of titanium or tantalum or a combination of both. The discharge is confined by a magnetic field, which causes the emitted electrons to travel in long spiral paths, under acceleration of voltages of 2,000 to 5,000 volts. Under these conditions, collisions occur between the energetic electrons and molecules of gas, which produce positively charged ions, and additional electrons, which continue the ionization process.

6.2 *Pumping Mechanisms*

The ions so created are attracted to the cathode and upon impacting a portion of them cause physical sputtering of the cathode material.

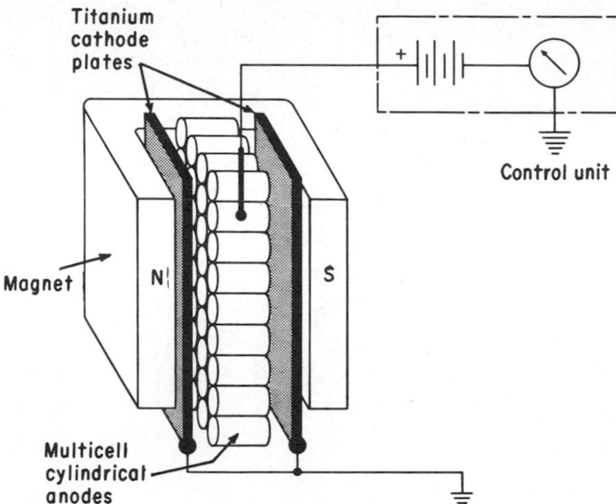

Fig. 6.1 Structure of a VacIon pump. (*Varian Vacuum Division, Palo Alto, Calif.*)

The sputtered material falls on both the cathode and anode surfaces, but preferentially at the points where these are closest together. The material so sputtered presents a clean, chemically active surface to the gas molecules present. The active gases oxygen and nitrogen combine chemically with the sputtered material to form stable oxides and nitrides, thus removing these gasses from the system (see Figs. 6.1 and 6.2).

Water vapor does not seem to be pumped directly but is first dissociated within the cell into hydrogen and oxygen, the latter element then being pumped by chemical combination. The dissociation process momentarily doubles the volume of gas, and at least partially accounts for the starting difficulties with ion pumps when connected to a system contaminated with moisture from the air.

Hydrogen is pumped by ion pumps at a higher rate than any other gas. However, the mechanism by which the hydrogen is pumped is

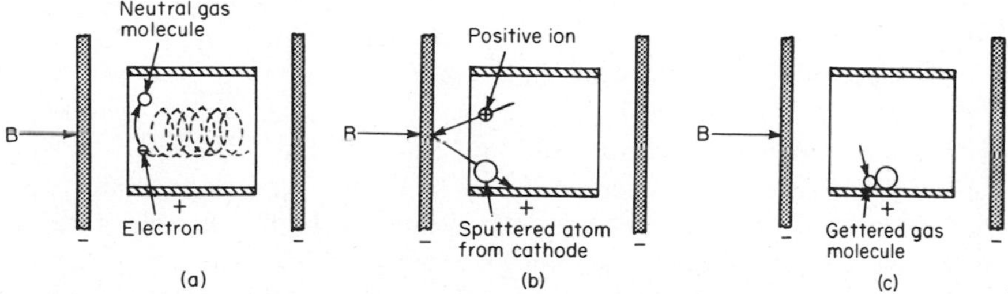

Fig. 6.2 VacIon pump: (*a*) gas ionization; (*b*) titanium sputtering; (*c*) chemical combination for active gases. (*Varian Vacuum Division, Palo Alto, Calif.*)

uncertain. Jepson and others have suggested that the hydrogen is first ionized, then accelerated to the cathode and buried therein, later diffusing inward into the bulk metal. However, titanium coated on the walls by thermal evaporation in a titanium sublimator pumps hydrogen with great rapidity without the presence of any electric field at all, and this fact argues against that theory. It seems more probable that the hydrogen first combines chemically with the titanium metal to form titanium hydride, afterwards, perhaps, diffusing inward into the bulk metal.

The noble gases of concern, argon and helium, form no compounds, hence can only be pumped by the burial method. The most annoying of the noble gases from the vacuum engineer's standpoint is argon, which is present to the extent of almost 1 percent in the atmosphere. Argon is a large and heavy atom, and does not seem to bury itself deeply on first impact with the cathode. In consequence, additional arriving atoms or molecules can easily dislodge the weakly bound argon. The net argon pumping speed in a standard diode-type pump is therefore very low, usually only 1 percent of the speed for air.

Several methods have been used in an effort to overcome this difficulty. One of these is the triode pump. This pump makes use of a double cathode arranged so that the incident ions strike the first cathode at an almost grazing angle of incidence. Under these conditions, the sputtering yield is at a maximum, and most of the sputtered material falls upon the secondary cathode. Argon ions or excited neutrals arriving at the secondary cathode are quickly buried under additional sputtered material. Such pumps seem to have a speed for argon of about 10 percent of the speed for air.

A second construction used to enhance the argon pumping speed employs a slotted cathode, in which a relatively heavy plate of material is slotted with comb-like serrations. This structure imparts to a diode pump the same grazing angle of incidence for approaching ions as in the triode pump and has much the same merits.

A third approach is to employ tantalum for one of the cathode plates. Speed for argon for this pump is found to be 10 to 15 percent of the speed for air. According to Jepson, a high proportion of the argon ions (mass 40) impinging on tantalum (mass 181) will be elastically reflected as excited neutrals, with most of their original energy. Many of the atoms will be backscattered and can thus reach the anode, where burial takes place and where incident ions cannot reach them to cause reemission. If this is the case, the relatively high pumping rate for argon of this type of pump becomes understandable.

The end effect of such a pump device is a pumping method which adds no contamination to the system in the form of oil molecules as do diffusion pumps but which still has a relatively high pumping speed.

Table 6.1 shows relative pumping speeds for standard diode ion pumps for various gases.

6.3 *Sublimation Pumps*

Ion pumps have enjoyed a considerable use for some years, but because of their relatively high cost per liter per second of pumping speed, they have been limited to small, clean systems. It remained for Hall to combine this phenomenon with the long-known phenomenon of gettering on freshly evaporated titanium in a way which greatly decreases the cost of the pump per liter per second of pumping speed. This method has now been adopted by other manufacturers and is available from several sources.

Basically, the system consists of an elongated pump section anterior to the actual ion pump itself and of larger diameter, in which a source of titanium vapor is placed. In most cases, this source consists of so-called "filaments" composed of a winding of tungsten heating element and a core of titanium wire. An external power supply is provided to heat the filament to a temperature at which the titanium begins to evaporate. At this point, it streams as a vapor to the walls of this chamber, forming a thin deposit upon the walls. Such deposits of fresh titanium have a great chemical affinity for all of the active gases, including hydrogen, nitrogen, and oxygen, and provide a very high pumping speed for these gases. Such deposits do not pump the noble gases—that is, helium and argon. In some very large pumps of this type, evaporation of the titanium is carried out by electron-beam heating of a titanium rod or by some other convenient means of providing a source of vapor.

It has been found that the amount of pumping which a given wall

TABLE 6.1 Pumping Speed for Various Gases Relative to Nitrogen

Gas	Pumping speed
Nitrogen	1.00
Oxygen	0.57
Argon, krypton, xenon	0.01 (0.06 for superpumps)
Hydrogen	2.7
Deuterium	1.9
Helium	0.10
Neon	0.05 (0.15 for superpumps)
Carbon dioxide, water vapor, carbon monoxide, etc.	1.0
Light hydrocarbons	0.90–1.60

SOURCE: Varian Vacuum Division, Palo Alto, Calif.

area can provide when supplied with fresh titanium in this fashion depends to a considerable degree upon the wall temperature. Thus the use of a liquid-nitrogen-cooled surface, or substrate, on which the titanium is deposited has been found to increase the amount of gas which may be pumped per gram of titanium evaporated by a factor varying from 2 to as high as 10 in various experiments. This is over and above such pumping as might be accomplished by a large liquid-nitrogen-cooled surface which, of course, has capacity for condensing water vapor. It would appear that the reason for this phenomenon is that the microscopic crystalline form of the film of titanium deposited on the substrate depends upon the temperature of the substrate, and where that temperature is maintained at a low value, a greater area is available for reaction with gas molecules than where the substrate is warm. For this reason, most pumps of this type employ at least water cooling of the substrate, and many of the larger ones employ liquid-nitrogen cooling of the substrate.

6.4 *Sublimation Filament Life*

The use of titanium sublimation for pumping of oxygen, nitrogen, and hydrogen from vacuum systems requires a preknowledge of pumping cycles, outgassing rates, and pressures of operation. At a pressure of 1×10^{-5} torr, a rather large amount of sublimation will be required to maintain maximum pumping speed. At a pressure of 1×10^{-10} torr, however, a total evaporation time of only 15 to 30 minutes per 24 hours will be required. One must therefore make sure in advance that sufficient filaments with external switching means and as many power supplies as will be required at the highest operating pressure are available and installed. Various systems may require from as few as 2 to as many as 48 installed filaments to handle the expected cycles and loads.

For large systems a bulk-type electron-beam sublimator is available which evaporates a bar of titanium 1¼ inches in diameter by 6 inches long and which provides sufficient material for many weeks of pumping at a speed in excess of 140,000 liters per second. Figure 6.3 shows such a system.

6.5 *Operating Characteristics*

Such pumps as those described above produce clean, dry (oil-free) vacuums and are capable of achieving good pumping speed down to at least 1×10^{-12} torr and probably lower. However, they can only be started when the pressure has been reduced by other means to values of 1×10^{-3} torr or below, since at higher pressures a glowing ionic discharge takes place which fills not only the pump but the whole cham-

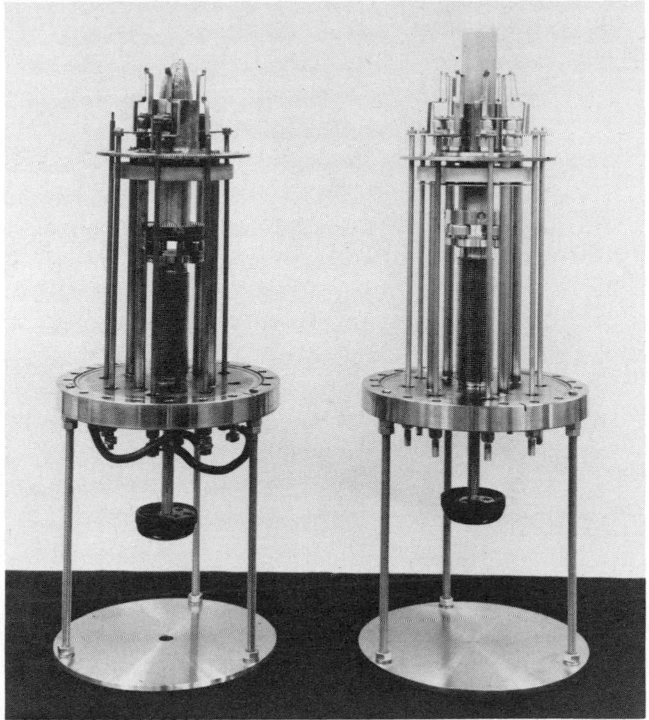

Fig. 6.3 Electron-beam bulk titanium sublimation for large ion pump. (*Ultek Division, The Perkin-Elmer Corp., Palo Alto, Calif.*)

ber with glowing gases. Because this is not confined to the active areas of the pump, great pumping speed does not result.

For these reasons, it is necessary that an ion-pumped system have available some means of roughing which will take the pressure down to at least 1 micron (1×10^{-3} torr) before the ion pump can be started. (Appropriate pumping systems are described in Sec. 4.4.) Good trapping should obviously be provided during the roughing, as the great advantage of the elimination of hydrocarbon contamination achieved with ionic pumping devices would be lost if the chamber were precontaminated with mechanical pump oil. Therefore, rather careful trapping arrangements must be provided, or pumps used that do not backstream.

6.6 *The Two-gauge Method of Determining Pump Speed*

This method of measuring pump speed makes use of the measured pressure drop over a known conductance to measure the gas being admitted to the system. An entrance tube, or sidearm, of considerable length is attached to the chamber as far as possible from the pump and at right angles to the pump entrance. Gas is admitted into the sidearm by means of an adjustable leak at a right angle to the tube axis. At a distance downstream from the leak of at least 4 tube diameters a gauge is inserted. Downstream an additional 10 diameters, a second gauge is attached. A third gauge is installed in the chamber to read pressure near the pump mouth. The conductance of the sidearm is now carefully computed using appropriate Clausing factors (see Chapter 19). The system is now pumped down to a pressure as near ultimate as possible and allowed to pump at this pressure at least overnight, all gauges being on and operating.

The pump is then shut off in order to calibrate the three gauges against each other. Equilibrium is considered to have been reached when three successive 5-minute readings show no change in the calibration or ratio of one gauge to the others. The correction of the gauge readings so obtained is used in all further runs. The speed-calibration run is made by restarting the pump, opening the leak, and reading the three gauges until stability is achieved. The inleak can be calculated by subtracting the corrected sidearm downstream gauge reading from the upstream gauge reading. The difference represents a pressure drop over the sidearm impedance $\left(\frac{1}{\text{conductance}}\right)$. A simple calculation gives the leak rate and, with the chamber pressure, the pumping rate of the pump. By varying the adjustable leak and taking new flow-meter readings, pumping speed at different pressures can be obtained, and a pump-speed curve constructed.

Several assumptions have been tacitly made in the above summary.

1. The pressure read by the chamber ion gauge is assumed to be correct. This is a shaky assumption unless the gauge has been calibrated by some accurate method (see Chapter 12). Since the calculated pump speed is inversely proportional to the chamber pressure as read on this gauge, the importance of using an accurate gauge is obvious.

2. A further assumption is made that outgassing in both the chamber and the sidearm is negligible. This will not generally be the case unless the chamber and sidearm have been held for 24 hours at a pressure 2 decades below the lowest pressure at which the speed run is to be made. Baking may be required to achieve this performance. Pressures close to the measuring pressure, during the preconditioning soak, may introduce large errors, especially if outgassing is taking place in the sidearm itself.

6.7 *Relative Pumping Speeds for Ion Pump and Sublimator*

Figure 6.4 shows pumping speeds for a nominal 50,000 liter per second pump employing both titanium-evaporation gettering and ion pumping. The tremendous gain achieved by the gettering technique in the evaporation section, especially at the lower pressures, indicates the efficiency of this method. It also indicates its inefficiency at the high-pressure end of the curve, where little is gained by the evaporative technique. This is primarily due to the fact that, at relatively high pressures, insufficient titanium is evaporated to maintain pumping for any appreciable period. The big gain is achieved in part because of the very low impedance into the large-diameter evaporative section, which, being of full-throat diameter without any impeding baffles, gives a very large gain. However, it should be pointed out that in a practical system it is usually necessary to interpose some form of baffling between the evaporative section and the chamber in order to prevent migration of titanium vapor into the work area, thus adversely affecting the work being carried out. This baffling, which need not be elaborate since it only has to close the line of sight, nevertheless does add some impedance to the passage of gas into the evaporative section and limits the maximum speed attainable accordingly.

The pumping speed of an ion-pumped system will vary greatly depending upon the kinds of gas molecules being pumped, since each gas

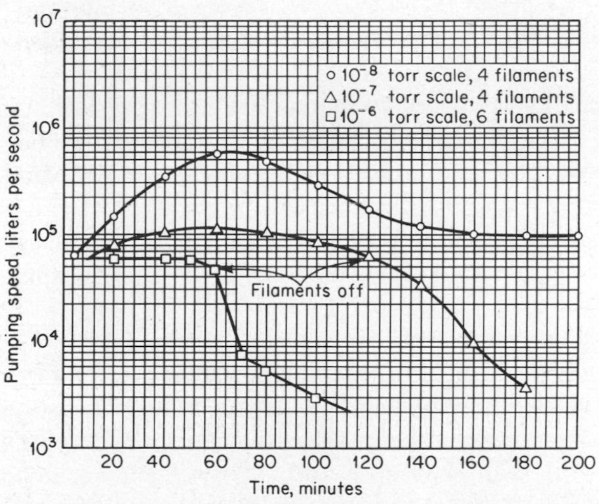

Fig. 6.4 Pumping-speed curves for nominal 50,000 liter per second ion pump with various filaments. (*Ultek Division, The Perkin-Elmer Corp., Palo Alto, Calif.*)

molecule is pumped as if it were the only type present, uninfluenced by the other gases. The relative pumping speeds for various gases in the straight-diode pump described up to this point will be found in Table 6.1. It will be noted that, whereas the pumping speed for hydrogen is very large, averaging 270 percent of the base speed of the pump, the pumping speed for helium and argon is very low, falling off (in the case of argon) to 1 percent of the speed for the active gases.

6.8 *Use of an Ion Pump as a Gauge*

The current passing through an ion pump is related to the pressure as in an ordinary Penning cold-cathode gauge. A current meter can therefore be calibrated in terms of pressure and will indicate pump pressure and, by inference, system pressure. This is only valid, however, if all the current being read is due truly to the Penning discharge. In practice, however, unless the pump is very clean and well baked, some leakage will take place across the ceramic insulators inside the cells. These small currents, while insignificant in terms of load, are quite sufficient to invalidate deductions regarding the chamber pressure. A separate ion gauge should therefore be employed in most cases.

6.9 *Selection of the Type of Pump*

The selection of the type of pump to be used on a system will normally be made not on a cost basis but on the basis of the degree of cleanliness required in the system. In general, a diffusion-pumped system will be somewhat cheaper than a system employing ion and titanium sublimation pumps; however, it will give a certain amount of hydrocarbon background, which for some purposes may not be permissible. In checking costs, we can assume that the roughing system will be required in either case, although in the case of the ion-pumped system it is frequently possible to use a common roughing system for several separate systems, since it is needed only during the startup period.

The cost of the diffusion pump and associated cold trap and elbow must be determined on the basis of net pumping speed at the chamber wall. The cost of the ion-pumped system, using titanium sublimation pumping as well, should be determined on the basis of its net pumping speed at the chamber wall with whatever form of rudimentary baffle may be required to prevent titanium from reaching the main chamber. This baffle is quite inexpensive, since no cooling is required. Under these conditions, the net cost of the system will generally turn out to be slightly in favor of the diffusion-pumped system, though not by a large amount. The diffusion-pumped system, however, can be used effectively at pressures ranging from 100 microns (1×10^{-1} torr) on down, whereas the ion-pumped system cannot be used effectively at pressures higher than 1×10^{-5} torr, since the pumping speed only reaches its normal value at approximately this pressure. Those systems which do not require extremely low pressures, therefore, are not particularly well suited to ionic pumping; whereas those that require extremely clean conditions at quite low pressures generally are advantageously pumped by ionic means.

chapter 7

Trouble-shooting, Cleaning, and Repairing Vacuum Pumps

7.1 *Introduction*

Before one can undertake to repair malfunctioning vacuum pumps, some care is necessary in determining where the malfunction lies and what steps are appropriate to correct it. In Fig. 2.1 we have shown a typical diffusion-pumped vacuum system employing a mechanical pump, a diffusion pump, cold trap, and valving. The first sign of trouble with pumping systems usually comes when it becomes impossible to pump the system down to the level normally obtained. The first step necessary is to determine what part of the system is causing the difficulty. For this reason, the sketch indicates valves separating each pump from succeeding pumps so that each can be isolated from other parts of the system.

The first step, therefore, might be to close the valve between the mechanical pump and the diffusion pump, having first turned off and cooled down the diffusion pump, and then to check the blankoff pressure achieved by the mechanical pump. If, under these conditions, the pressure between the cutoff valve and the mechanical pump does not fall to a value appropriate for the type of mechanical pump being used, the source of trouble must be in the mechanical pump. For single-stage mechanical pumps, the blankoff pressure should be somewhere between 10 and 25 microns (1 to 2.5×10^{-2} torr) as measured by a thermocouple gauge; for a two-stage pump, the pressure should be somewhere between 1 and 10 microns (1×10^{-3} to 1×10^{-2} torr) as measured by a thermocouple gauge capable of accurately reading pressures at these levels. A word of warning should be inserted here: Most commercial thermocouple or Penning gauges do not, in fact, read pressures accurately below 25 microns (2.5×10^{-2} torr), giving a steady reading at this value even though the pressure falls below this point.

If it is determined that the mechanical pump is operating properly, the valve between the mechanical pump and the diffusion pump should be opened, the diffusion pump should be heated while the valve separating it from the chamber remains closed, and the pressure above the cold trap but between it and the valve should be read by means of a suitable ionization gauge. Under these conditions, with the cold trap cooled by liquid nitrogen, the pressure should fall rather quickly after the pump once stabilizes itself to values in the neighborhood of 1×10^{-6} or 1×10^{-7} torr. If the cold trap is not cooled, then equivalent values would be approximately 1×10^{-5} torr. If these values are not attained although the mechanical backing pump is operating properly, then the trouble must be in the diffusion pump.

In the rather common case where both of these pumps are found to be operating properly, we must then seek the cause of the trouble in the chamber being evacuated, and a rigorous course of leak checking is required to locate the leak and correct it.

7.2 *Cleaning and Repairing Mechanical Pumps*

The most common causes of trouble with mechanical pumps are as follows:

1. ***Motor Trouble*** This may be due to single-phasing a three-phase motor or to slippage of the belt connecting the motor and the pump. Either of these will, of course, cause the pump to run at less than its normal speed and, consequently, degrade its performance. The remedies are so obvious as to require no further comment.

2. ***Contaminated or Insufficient Pump Oil*** The quantity of pump oil within the reservoirs may readily be checked by removing the usual cap or by observing a gauge where this is provided; if the quantity is indeed low, oil must be added to bring it up to the proper level. Occasionally, it will happen that the oil will be found to be too high due to some pressure burst in the chamber having forced diffusion-pump fluids over into the mechanical-pump side. Where this has happened, the oil must be drained and the pump refilled since most diffusion-pump oil does not have satisfactory characteristics for operation of a mechanical pump.

A more common cause of pump-oil difficulties lies in contamination of the oil, usually by water vapor. This is sometimes discernible by a milky appearance in the oil due to emulsified water; but the absence of such appearance is no guarantee that the oil is not, in fact, contaminated. If time is available and if there is no reason to suspect excessive water migration into the mechanical pump, cleanup of the oil can usually be accomplished by running the pump overnight with the ballast valve slightly open and with the pump blanked off from the system. Under these conditions, the pump oil will frequently free itself of water vapor if the quantity involved is not too great.

If time is not available for this type of cleanup action, or if there is reason to suppose that large quantities of water vapor have been permitted to enter the mechanical pump, then draining the oil and refilling with fresh oil is mandatory. Frequently, a single drain and refill will not completely eliminate the difficulty; the operation may have to be repeated as many as four times where very large amounts of water vapor, due to a failure within the system, have been allowed to enter the pump oil.

If pump oil is bought in large quantities and used in smaller amounts from time to time, it is frequently found that the oil within the large container has become contaminated during storage. It is then necessary to run the pump overnight, blanked off and with the ballast valve open, in order to condition the new oil drawn from that container to a point where it will pump down to the required pressure.

In general, it is worthwhile to attempt correction of pump operation by oil changes, since if this is not done, the only other option involves tearing down the pump, which is time-consuming and expensive. However, if pump cleanup by means of oil changes or blank-off operation does not solve the difficulty, then a pump teardown will be required.

3. *Mechanical Difficulties* The most frequent causes of mechanical trouble with pumps of either the rotary or the rotary vane type are leakage by the shaft seal or improper valve operation. Either or both of these conditions will require replacement of the affected parts and readjustment of the assembly. In addition, once the pump is torn down, all parts should be thoroughly cleaned in acetone and all accumulated carbon or dirt that may be present very carefully removed.

Reassembly of a mechanical pump after teardown is a job requiring considerable skill on the part of the mechanic, and it is not usually worthwhile to attempt this oneself unless prior experience has been had in this field. There have been several cases where expert mechanics highly skilled in refrigeration compressors have had to spend as much as a week to ten days in reassembling a pump after teardown before it could be pumped down to its required blankoff pressure. If skilled help is not available in this field, it is undoubtedly more economical to return the pump for reworking to one of the numerous repair shops operated by pump manufacturers. This is particularly true in the case of the rotary vane pumps, where reassembly of the rotor and adjustment of clearances is especially difficult, although it applies also to the rotary piston-type pumps to some extent. Once skill is acquired in this field, however, the disassembly, cleaning, and reassembly of any of the various sizes of mechanical pumps should not require more than two days at the most, assuming the necessary seals, spare parts, etc., are available.

7.3 *Cleaning and Repairing Diffusion Pumps*

The diffusion pump, unlike the mechanical pump, has no moving parts, and disassembly and cleaning are therefore somewhat simpler. The principal causes of trouble in diffusion pumps are as follows:

1. One or more of the several heaters employed may be burned out. The remedy, of course, is obvious.

2. The pump working fluid or oil may have been cracked by being exposed to air while heated. Such exposure will cause breakdown of the long chain molecules of which the oil is composed in a very few seconds, so that a sudden pressure burst, in the case of pumps that are being operated with hydrocarbon-type fluids, almost invariably results in cracking of the oil, with the result that the vapor pressure of some of the cracking products becomes too high to permit continued use in a diffusion pump.

It is seldom worthwhile to change the oil in the diffusion pump without at the same time removing the pump from the system and cleaning it thoroughly with vapor degreasers, acetone, or other cleansers. When one carries out this process, it is frequently found that relatively large deposits of tar and hard carbonaceous materials have formed in and around the heater surfaces. These must be removed before the pump is returned to service, as they act as insulators and cause the heaters to operate at a higher temperature than normal, resulting in additional cracking of the oil adjacent to these hot spots.

Removal of these tenacious deposits is not an easy matter since they are not readily soluble in any of the available solvents. They can be removed by means of steel wool, but for some pumps it becomes very difficult to reach down to the bottom of the boiler in order to work with the steel wool, with the result that it has to be attached to the end of a long rod which greatly increases the difficulty of removing the tenacious deposits. The most successful method that has been developed for this purpose involves a liquid grit blast in which small grit particles are circulated against the contaminated areas by a fluid, generally water. This method removes the carbonaceous deposits very quickly and enables the inside of the pump to be cleaned up in a very short time. Before reassembling the pump and refilling it with its working fluid, it should be thoroughly washed and cleaned with acetone or alcohol in order to remove all traces of the water resulting from the cleaning action carried out earlier. All heating elements in the heating section should be checked for continuity and the pump then refilled with the proper quantity of the proper grade of oil.

It is appropriate at this point to mention that diffusion pumps are quite sensitive both to the quantity of oil used and to the grade of oil used; it is essential in all cases that the manufacturer's recommendations be closely adhered to. Where changes from organic fluids to silicone fluids are to be made, the heat input should be readjusted to correspond to the needs of the new fluid since, in general, various fluids require various amounts of heat input to secure proper functioning of the pump. The use of silicone fluids is becoming much more widespread, in spite of their higher cost, due to their greater stability when subjected to the type of pressure burst occasioned by failure of some component of the vacuum system. These fluids, while not completely immune to the cracking reaction which causes destruction of the diffusion-pump fluid, are much more resistant to damage and will frequently withstand several such pressure bursts before they crack sufficiently to cause difficulties.

It is unusual to find diffusion-pump fluids seriously contaminated by materials coming from the vacuum chamber, since most such materials have sufficiently high vapor pressures that they are removed from the diffusion pumps and passed on to the mechanical pumps where they may or may not cause troubles in the mechanical-pump fluid. However, there are some exceptions to this condition, principally when the influent materials are such as to react chemically with the oil molecules. In this case, early destruction of the pumping ability of the diffusion pump will ensue. Where the influent gas being pumped is pure oxygen, reaction with pump oils can be serious and even dangerous, since the organic fluids are sufficiently oxidizable so that combustion and explosion may occur if the oxygen concentration reaches a high value. Where such conditions are apt to occur, a relatively inert silicone material, such as DC-705, should be used.

The various pump oils mentioned above have, in addition to differences in their stability under long-term use, differences in their vapor pressures. This can be serious, since the vapor pressure of the oil at the temperature of the trapping medium will determine the number of oil molecules that migrate back into the system and hence degrade the system vacuum. Generally, there is very little difference between the various fluids when used with a highly efficient liquid-nitrogen-cooled cold trap; but where water- or Freon-cooled baffles are used alone, the ultimate pressure attainable will be directly proportional to the vapor pressure of the fluid employed. For this reason, it may well be that for some systems where liquid-nitrogen trapping of good design is not employed, it will turn out to be better to use a very expensive oil such as Convalux-10 or DC-705, since these will give lower chamber pressures at ultimate than the more volatile hydrocarbon oils which cost so much less.

In reassembling the jets and stack of the diffusion pump, it is quite important that these be placed in exactly the position in which they were originally designed to be, as slight changes in the positioning of the stack will change in a major way the throughput and ultimate pressure attainable with a given diffusion pump. Usually, guide pins or projections of some variety are provided at the bottom of the diffusion-pump barrel which serve to locate the stack accurately and perpendicularly. However, if these are not snugly fitting, it is still possible to assemble the pump in such a way that the stack is positioned out of the vertical and is not centrally located within the barrel of the pump. In this case, pumping performance will suffer markedly. Also, where an ejector stage is employed in the foreline (to permit use of higher fore-pressures), it is important that the vapor passageway in the skirt of the stack be properly oriented if full throughput is to be achieved.

Some pumps are quite sensitive to the quantities of cooling water used and to the temperature of this cooling water. It is suggested that some experimentation be carried out to determine the pump performance as a function of cooling-water temperature and flow rate, since these can affect the total pumping speed and the ultimate pressure thereby attained. Where automatic flow-control valves are used, one should be careful to provide some form of emergency alarm which will turn off the pump power in the event of water failure or low flow rates caused by partial clogging of the control valve. This has been a rather frequent cause of heater burnout and pump failure in systems where automatic control of water temperature is employed, although if this functions properly, improved performance should result.

3. Contamination of diffusion-pump oil: It frequently happens that during pumpdown cycles conditions arise in which oil from the mechanical pump is backstreamed into the diffusion-pump boiler. Where this

happens, the diffusion-pump oil will, of course, have much too high a vapor pressure and must be drained and the pump cleaned and refilled. A cold trap between the mechanical pump and the diffusion pump is desirable, but is not frequently provided because of the expense both for the trap and for the liquid nitrogen to keep it operating. If operation is carried out in the proper manner, no trouble should arise from this source; but where pressure bursts or some misoperation occurs, this type of failure can result.

7.4 *Cleaning and Repairing Ion Pumps*

Ion pumps have no moving parts and no working fluid, so are somewhat freer from difficulties in operation than either mechanical pumps or diffusion pumps. However, they do present some problems. The most frequent of these is the shorting out of some of the insulators in the pump body due to deposition of cracked organic materials which have come from the chamber being pumped or possibly from metallic vapors from the sputtering action of the pump where shields were not properly arranged. When this happens, one or more elements will of course be inoperable and frequently the power supply will shut down. The remedy is to disassemble the pump, remove the elements, clean—or better yet, replace—all the insulating ceramics that are employed in the high-voltage system, clean the anode and cathode structures by pickling, and reassemble. After reassembly, the pump will be somewhat contaminated due to water vapor and must be given a thorough bakeout, either by internal bakeout heaters (if provided) or by an external heating means, preferably at temperatures of 300°C or higher. The pump magnets must be removed during high-temperature bakeout to avoid damage, especially if external heating means are employed. During bakeout, the ion pump must be maintained at a pressure of 100 microns (1×10^{-1} torr) or below by some form of external pumping system, well trapped to avoid backstreaming of hydrocarbon materials.

The first noticeable symptom of a badly contaminated pump is usually difficulty in starting. A vigorous blue discharge takes place not only within the pump body but throughout the whole chamber; this may continue for several hours, with severe heating of the pump. Some of this effect is always evident when the ion pump is first turned on; but if the pump is clean, it should recede and be confined within the pump cells within 15 to 30 minutes of the time the pump is energized. True pumping action normally begins when the supply voltage has risen to approximately 1,100 volts, after which the pressure will fall rapidly to 1×10^{-5} torr or below.

The energetic blue discharge (ionic discharge) can be quite damaging to various parts of the system. Hot-filament ion gauges should be turned off during the discharge since filament burnout is very probable if they are left on. The insulation on electrical leads (other than glass or quartz) will frequently be stripped off by the energetic ions.

The discharge just referred to can be confined within the pump body by the use of a ½-inch mesh stainless steel screen, placed in the pump throat and well grounded. Such a coarse mesh screen offers little impedance to the flow of gases in the molecular flow regime, but effectively confines the glow discharge and consequently accelerates the pump startup.

Cold-cathode gauges of the General Electric or Kreisman types do not seem to suffer damage of a permanent nature due to glow-discharge effects, although the discharge will cause rapid fluctuation of the meter.

Titanium sublimation systems used in connection with ion pumps should normally not be energized until the pressure has been reduced to 1×10^{-5} torr or below, since very little is gained in speed at higher pressures. When the filament is first turned on after being exposed to air, a pressure rise of some considerable amount, due to filament outgassing, is usually noted. This can be largely avoided by energizing the filaments during the roughing stage at a current value just below that at which sublimation begins. This allows the outgassing to be removed from the system by the roughing pump before the ion pump is energized and the roughing pump disconnected The outgassing occurring during the first few sublimation cycles is thereby greatly reduced.

Starting of ion pumps can be made much easier by maintaining the pump at approximately 150°F during the period when the system is exposed to air. This slightly elevated temperature prevents sorption of water vapor or other contaminants by the pump elements during exposure, thus rendering starting much easier. If a slightly elevated temperature can also be maintained on the chamber proper during the period when it is open to air, the problem of starting and pumpdown can be further reduced. This may not be possible or convenient in some cases, but the pump-temperature method can usually be carried out conveniently. Care must of course be exercised to keep the pump temperature below 200°F to avoid oxidation of the pump elements while the system is exposed to air.

Where the system must be open for prolonged periods so that maintenance of heating is not practical, the application of 200°F bakeout during roughing to the pump alone, or to the pump and system as well, can be helpful. Pressure in the system should be maintained at approximately 100 microns (1×10^{-1} torr) in order to achieve viscous flow, thus sweeping water vapor from the system. Thirty minutes of such purging is very effective in reducing the amount of adsorbed water on the internal parts of the system. A full bakeout at 300°C is obviously much better, but may not be permissible.

The frequency of cleanup and/or bakeout operations on ion pumps will depend entirely upon the use to which they are put and the amount of contamination introduced into the pumps from the chambers to which they are attached. Very long periods of operation with such pumps can be anticipated from clean systems, but rather frequent cleaning may be required where organic materials (hydrocarbons) are released within the chamber during vacuum operation and permitted to reach the pump.

The actual life of the pump elements is determined by the number of hours of operation and by the pressures at which the pump has operated. The period, under normal conditions, is extremely long; for systems operating at 1×10^{-6} torr or lower, it is probably of several years' duration. However, cleaning within this period may be required.

7.5 *Cleaning Ion-pump Elements*

Since the materials used in various makes of ion pumps vary widely, no precise statement can be made as to the type of pickling bath to be employed. Generally, however, hydrochloric acid solutions and mixtures of sulfuric and nitric acids are used. An appropriate solution for the metals to be cleaned must be used. In the proper solution, a pickling time of very few minutes will be sufficient. Longer times may result in complete destruction of the thin metals used in construction of egg-crate anodes and flat-plate cathodes.

Insulators are difficult to clean, but tumbling and glass-bead shotblasting have been used with reasonable success. Some breakage will occur, and some insulators will not clean up satisfactorily. Generally, replacement of from 10 to 25 percent of the beads will be necessary. Metal parts will usually survive four to five cleanings before replacement is necessary.

Sublimation coats the substrates with a thin layer of titanium, which turns grayish blue as it combines with oxygen and nitrogen. If this process is carried on long enough, flaking of the deposits eventually occurs; these flakes may short ion-pump elements or possibly find their way into the chamber. To avoid this effect, the substrate surfaces should be cleaned when the system is opened for cleanup of the ion-pump insulators. If the structure permits, pickling is indicated. In the more usual case where this is not feasible, buffing with a motor or pneumatically driven stainless steel wire-brush wheel can be used advantageously.

Normally, such cleaning should not be necessary more than once a year, unless severe contamination has occurred, such as an oil leak, in which case it must be done immediately.

7.6 *Reliability*

All communities suffer occasional, but infrequent, power outages. In a diffusion-pumped system, such failures cause loss of backing pressure, loss of pump heater power, a closure of any isolation valves provided with spring or pneumatically operated mechanisms, and gradual loss of cooling water, cryogenics for cold traps, etc. Theoretically, the automatic closure of the isolating valves should prevent any oil contamination from reaching the chamber. Vacuum will be partially or totally lost within the chamber, which may or may not adversely affect the work being done within the chamber.

In practice, such an occurrence is seldom without some adverse effects. In two such cases personally known to the author, such adverse effects *did* occur, causing serious and very expensive contamination of material within the chamber. In another case, where contamination did not occur, the interruption of the test made it necessary to repeat three weeks of testing. Where the contamination occurred, a long chamber outage was required while the chamber was partially disassembled and thoroughly cleaned.

In ion-pumped chambers employing titanium sublimation as well, no contamination occurred due to shutdown except for a slight rise in the methane content of the chamber gases. The pressure rise due to an outage was limited to about 2 decades. When power was restored, the chamber pumped down again in about 10 minutes. This is a real advantage of the ion-pumped system.

7.7 *Mechanical Booster and Turbine Pumps*

Both of these types of pumps are relatively high-rotating-speed devices, indeed extremely high-speed in the case of the molecular-drag or turbine pump. When difficulties arise, if they are of any but the most simple nature, it is advisable to return such pumps to the manufacturer for servicing. The problems of assembly, balancing, and adjustment of the very close clearances involved are such that without special tools and fixtures success is unlikely.

Such maintenance on pumps of this type should be quite infrequent and, in fact, occasioned only when mistreatment of some kind has occurred, since the pumps have no rubbing parts and should have a very long life if no untoward accident happens to them. Bearings are probably the most important elements requiring service in such pumps; but, as mentioned above, the very close tolerances employed make it advisable that such repairs be done by those possessing the necessary jigs and fixtures to assure proper reassembly and dynamic balancing.

chapter 8

Cryogenics in Vacuum Systems

8.1 *Introduction*

Cryogenics is a term derived from a Greek word meaning "icy cold." It has acquired a connection with the techniques of dealing with extremely low temperatures as produced in or by liquefied gases such as liquid nitrogen, liquid oxygen, liquid hydrogen, liquid helium, and very cold gaseous helium.

Such materials are used in vacuum systems for cooling cold traps, for the condensation of materials either coming from pumps or from systems being evacuated, and for condensing gas molecules which may be frozen at the temperature of the cryogenically cooled surfaces. Some knowledge of the properties of liquefied gases is therefore necessary in connection with vacuum systems.

8.2 *Storage of Cryogenic Fluids*

Cryogenic fluids at low temperatures are stored in dewars which consist of glass or metal vessels of a double-wall construction, the space between the two walls being evacuated in order to lower the heat transmission through the walls into the cold liquefied gas in the vessel.

Small dewars used for the transportation of some of the cryogenic fluids are frequently made of glass with an evacuated space between the double walls after the fashion of thermos bottles. However, larger vessels are invariably fabricated of stainless steel welded by the tungsten–inert-gas arc-welding process to achieve gas-tight closures. In general, all storage of liquefied gases is carried out at atmospheric pressure or at pressures of a relatively low value, utilizing some form of blowoff valve to release sufficient gas from time to time to maintain the temperature of the remaining fluid below its boiling point at the pressure selected through the heat of vaporization of the portion vented.

8.3 *Cryogenic Lines*

Lines carrying liquid nitrogen or cold helium around laboratories to cold traps and the like, if of a permanent nature, should be insulated by means of vacuum jacketing, employing superinsulation within the jacket in the case of helium and a properly evacuated open jacket in the case of liquid nitrogen. Such jacketed lines are rather expensive to fabricate, since, in general, both the inner and outer tubes must be of stainless steel and must be welded by the tungsten–inert-gas process to perfect gas tightness. Provision for the differential expansion of the inner and outer lines must be made by means of bellows employed usually in the outer jacket in order to compensate for the change in dimensions of the inner tube when it is filled with the cryogenic fluid. Such lines will probably have an installed cost ranging from \$50 to \$100 per linear foot, which has deterred some from making use of this construction.

Pressures in the distributing-line jacket should be sufficiently low to prevent convection losses from the inner tube to the outer tube, and experience indicates that this requires a vacuum of approximately 1×10^{-5} torr. This can be achieved by pumping the systems with small diffusion pumps which may be turned off after the system has been pumped down, since leakage will be small. However, outgassing from the surfaces of the inner and outer tubes and from the superinsulation layers, if used, will still take place; and over a period of months, the vacuum in the jacket will usually deteriorate. Gauges should be provided which will enable this to be checked from time to time, so that when the vacuum has deteriorated to a point increasing the leakage of heat into the fluid to an unreasonable extent, pumping can be carried out again to restore the line to its original condition. In a well-constructed line, repumping probably will not need to be done more than once every six months. Appendage ion pumps are sometimes used.

These use very small amounts of power and can be left permanently in the "on" condition to pump any leakage or outgassing which may occur. They have proved very useful for large cryogenic systems.

The same need for repumping exists on large storage dewars which from time to time must have the vacuum renewed within the insulating jacket.

8.4 *The Use of Cryogenic Fluids as Pumping Mediums*

It has long been known that the cold traps cooled by liquid nitrogen, employed to prevent backstreaming of oil from the diffusion pumps, act also as water-vapor traps, pumping the water vapor coming from the test vessel by condensing the water on the cooled surfaces. This effect is quite efficient in reducing pumpdown time, since the greater part of the gas that must be evacuated from chambers at low pressures is water vapor, which can be pumped much more rapidly by cold surfaces than by diffusion pumps.

As very large vacuum chambers came into use for space simulation work for the testing of satellites and space vehicles, the cost problem of pumping these large chambers became acute. If all of the pumping were done by diffusion pumps, the entire surface area of the vessel became covered with pumps, each with a cryogenic elbow and trap, resulting in very high costs—considerable power costs for operating the pumps plus the high cost of providing sufficient strength to the vessel to permit such a large number of penetrations and to provide backup pumping for all of the pumps.

To solve this problem, use was made of cryogenic pumping within the chambers. Liquid-nitrogen shrouds alone, installed within the chambers, provide a very great increase in pumping speed for those gases, principally water vapor, condensible upon the cryogenic surfaces at liquid-nitrogen temperatures. This aided the evacuation of the chambers, but had distinct limitations since oxygen is pumped on liquid-nitrogen surfaces only very ineffectively—and nitrogen, hydrogen, and the other gases, not at all. Figure 8.1 shows the vapor pressure of various gases at various temperatures. As can be seen from inspection of the curve, even those gases which can be pumped on liquid-nitrogen-cooled surfaces (with the exception of water vapor) tend to reevaporate from these surfaces as the pressure goes down to some of the values that are now used.

In order to improve the situation, secondary pumping surfaces cooled by helium gas at a very low temperature are employed. The best results in this connection would be achieved by the use of liquid helium. However, the difficulties inherent in the use of liquefied helium have resulted

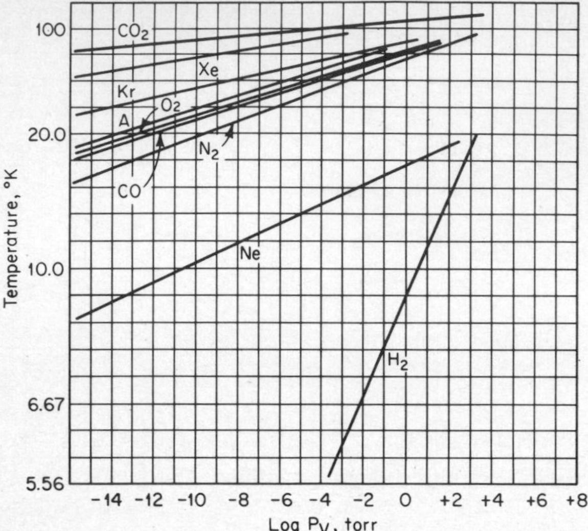

Fig. 8.1 Vapor pressure versus reciprocal temperature for solid phase of various gases. (*Pittsburgh–Des Moines Steel Co.*)

instead in the use of very cold helium gas circulated at temperatures of 18 to 20°K or below. This gas can be produced in a helium refrigerator system with reasonably good economy, circulated through the pumping panels inside the chamber, then returned to the system for recooling without loss of the quite expensive helium gas. In order to lower the heat load on the helium surfaces to a minimum, these are shielded from sight either of the chamber walls or of the work piece inside the chamber by liquid-nitrogen-cooled panels.

With this arrangement it can be seen that all of the gases expected to be present in vacuum systems may readily be condensed on the surfaces of plates operated at temperatures below 20°K with the exception of hydrogen, neon, and helium. These gases must therefore be removed by means of primary pumping systems of either the diffusion- or ion-pump type. The greater part of the gas present, however, may be cryogenically pumped by the cooled surfaces.

8.5 *Pumping Speed of Cryogenic Surfaces*

The actual pumping speed of a cryogenically cooled surface will depend upon the number of molecules which impinge upon it (which is, of course, a function of pressure and of temperature, which controls the speed of motion of the molecules), of panel geometry, and upon the probability that a molecule impinging upon the cooled surface will stick to the surface. Numerous tests have been run to measure or attempt to measure the probability of a molecule which impinges on a cooled surface sticking to that surface. It would appear that, for those gases which are capable of being solidified at the temperature of the cooled surface, the sticking probability for those molecules actually impinging on the surface is better than 90 percent.

It is important to notice, however, that as the pressure goes down some of the solid molecules frozen at higher pressures will sublime and reevaporate unless the temperature of the cooled surface is maintained below the sublimation point. Thus while nitrogen can be collected on cooled surfaces at temperatures as high as 78°K, these molecules will reevaporate as the pressure drops; therefore, to hold such species on the cooled surface at 1×10^{-10} torr, a cold-surface temperature of lower than 18°K is required.

Both calculations and speed measurements on fully exposed liquid-helium-cooled surfaces have been made. These indicate that for such fully exposed surfaces, pumping speeds of the order of 10,000 liters per second per square foot of cooled surface can be expected. However, this theoretical speed cannot usually be obtained in practice for the reasons outlined in the following section.

8.6 *Cryopumping Calculations*

The rate r_i at which molecules of gas in a high vacuum, characterized by free molecular flow, strike a unit condensing surface is

$$r_i = n \frac{V}{4} \tag{8.1}$$

in which

$$V = \left(\frac{8RT_i}{\pi M}\right)^{1/2} \quad \text{for an ideal gas}$$

and

$$n = \frac{MP_i}{RT_i}$$

where V = mean molecular velocity
n = number of molecules per unit volume
R = universal gas constant
P_i = incident gas pressure
M = molecular weight
T_i = incident temperature

Therefore

$$r_i = \frac{P_i}{(2\pi R)^{1/2}} \left(\frac{M}{T_i}\right)^{1/2} \tag{8.2}$$

The rate r_c at which condensed molecules evaporate from the cryosurface is similarly derived for the vapor pressure P_c and temperature T_c prevailing at the cryosurface, or

$$r_c = \frac{P_c}{(2\pi R)^{1/2}} \left(\frac{M}{T_c}\right)^{1/2} \tag{8.3}$$

The net rate equals

$$r_i - r_c = \frac{P_i}{(2\pi R)^{1/2}} \left(\frac{M}{T_i}\right)^{1/2} \left[1 - \frac{P_c}{P_i}\left(\frac{T_c}{T_i}\right)^{1/2}\right] \tag{8.4}$$

By dividing this mass rate by the gas density, the theoretical cryopumping speed S_{th} equals:

$$S_{th} = \left(\frac{T_i R}{2\pi M}\right)^{1/2} \left[1 - \frac{P_c}{P_i}\left(\frac{T_i}{T_c}\right)^{1/2}\right] \tag{8.5}$$

For nitrogen: $P_c \approx 1 \times 10^{-10}$ torr at 20°K
For oxygen: $P_c \approx 1 \times 10^{-14}$ torr at 20°K

The ratio

$$\frac{P_c}{P_i}\left(\frac{T_i}{T_c}\right)^{1/2}$$

can be neglected for pressure levels several magnitudes higher than these vapor pressures.

In other words, the pumping speed is nearly independent of pressure until the vacuum level reaches an order of magnitude close to the vapor pressure. The actual pumping speed S_a equals:

$$S_a = S_{th} G \tag{8.6}$$

The factor G is the capture probability or sticking coefficient and is defined by the net number of molecules trapped divided by the number of total incident molecules. Capture probability is a function of the geometry of the array and the proximity of the test object. It can be determined by assuming a cosine distribution of molecules reflected from the thermal shroud and cryoshield.

For an infinite empty test chamber with a cryoarray in which panel width equals panel spacing, 50 percent of the incident molecules will reach the shroud; and of these, 50 percent will be reflected past the cryosurface. Therefore, the capture probability for such an array is

0.25. It can be shown that this is the optimum case. For other width-to-spacing ratios, G will be less than 0.25.

If we base S_{th} on nitrogen (S_N), the pumping speed becomes:

$$S_a = GS_N \left(\frac{28}{M}\right)^{1/2} \tag{8.7}$$

$$S_N = \left(\frac{T_i R}{2\pi M}\right)^{1/2} \tag{8.8}$$

At 300°K:

$$S_N = \left(\frac{300 \times 8.31 \times 10^7}{2\pi \times 28}\right)^{1/2} = 1.19 \times 10^4 \text{ cu cm/sec}$$

Expressed in liters per second per square foot of array area,

$$S_N = \frac{1.19 \times 10^4 \times 30.48^2}{1000} = 10{,}800 \text{ liters/(sec)(sq ft)}$$

and pumping speed for any gas capable of condensation becomes

$$S_a = 10{,}800 \times G \times \left(\frac{28}{M}\right)^{1/2} \quad \text{liters/(sec)(sq ft)} \tag{8.9}$$

If the incident molecular temperature deviates much from 300°K, a factor $(T/300)^{1/2}$ should be applied.

The factor G above has been called the "sticking factor" by other authors. It varies not only with frost accumulation, but with the vapor pressure of the solid form of the gas being pumped, the angle of incidence, and perhaps with various surface roughness effects. It is *not* well established except for definite cases.

8.7 *Cryogenic Arrays*

If a helium-cooled surface at temperatures below 20°K is freely exposed inside a chamber where the outer shell is at room temperature, and if a vehicle or sample being tested in the chamber is also at room temperature or warmer, the consumption of coolant becomes so high as to be impractical. Practical cryogenic arrays, therefore, involve the shielding of the helium-cooled surfaces from the warm chamber walls by means of a liquid-nitrogen shroud maintained at 100°K or below. Such shrouds are interposed between the wall and the helium-cooled surfaces in a substantially continuous manner.

In addition, a secondary shroud or an extension of the primary shroud is so arranged that direct heat from the specimen inside the chamber cannot reach the helium-cooled surfaces without first being reflected

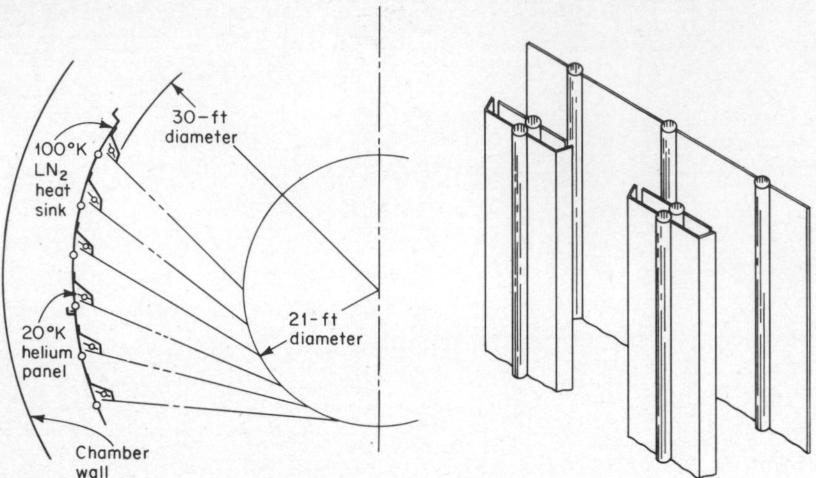

Fig. 8.2 Santeler-type cryopanel array.

Fig. 8.3 Venetian-blind-type cryopanel array.

from the nitrogen-cooled surface. Typical arrays are shown in Figs. 8.2 and 8.3. The array shown in Fig. 8.2 is that developed by Santeler for the General Electric Company, generally known by his name. This array affords 100 percent protection for the helium-cooled surfaces only if the diameter of the array is considerably larger than the diameter of the parts being tested. If the diameter of the part being tested exceeds 70 percent of the shroud diameter, it is apparent that heat radiation can reach the helium-cooled surfaces from the extreme edges of the specimen. Such arrays therefore are wasteful of chamber volume or result in the consumption of additional quantities of the helium coolant above that which might otherwise be necessary. The great advantage of the Santeler array, however, is that its design offers less impedance to gas molecules which are to be pumped by the helium-cooled surfaces than other arrays. Such an array has an efficiency of better than 60 percent; that is, the pumping speed of the helium-cooled surfaces can be approximately 60 percent of the theoretical values given in the preceding section.

The array shown in Fig. 8.3, consisting of overlapping panels of nitrogen-cooled surfaces maintained at 100°K or below, produces much better economy in the consumption of the helium coolant. However, its pumping efficiency is not as good, since the impedance offered to migrating molecules on the way to the helium-cooled surfaces is much greater, giving an overall efficiency for this type of array of 40 percent or less.

In practice, one is forced to compromise between these two methods, making use of an array which, for the particular size vessel to be used and the size specimens to be run therein, offers the greatest advantages in pumping speed as well as in economy of helium refrigerant.

8.8 *Surface Treatment of Arrays*

In order to cause the liquid-nitrogen-cooled panels to absorb all the heat originating from the specimen being tested as efficiently as possible and also to prevent reflection of light from solar simulators from these arrays back to the vehicle being tested, it is customary to blacken the internal surfaces of the arrays by some means which will raise their emissivity close to that of a black body. Black epoxy paints of a special formulation have been developed which when properly applied will give emissivity values well in excess of 0.9 when new. Surfaces of the nitrogen-cooled panels facing outward toward the chamber shell are normally left bright, as are those surfaces which face the helium-cooled array only. The purpose of this treatment is to lower the emissivity of these surfaces to a minimum value, preferably less than 0.1, so that these panels absorb as little as possible of the heat coming from the outer shell and radiate as little as possible toward the helium-cooled array.

Unfortunately, these treatments of the shroud surfaces, especially the use of high-emissivity paints, render it impossible to use bakeout procedures in chambers containing shrouds. If the temperature is raised to a point which will thoroughly drive off gas molecules, the coating is also driven off or reduced in such a fashion that it no longer performs the intended function. The paints, unfortunately, appear black (at least in part) because of the very large surface area they expose; this same feature permits them to trap large quantities of water vapor during exposure to the atmosphere. This water vapor is then given off copiously during evacuation and presents considerable difficulty in the attainment of low pressures. This effect can be eliminated by cooling these panels with liquid nitrogen which effectively freezes all the water vapor present on the surfaces and prevents its release into the vacuum chamber proper. When so cooled, the outgassing of the shrouds becomes a negligible factor.

8.9 *Liquid-nitrogen Systems*

Liquid-nitrogen systems can be of either the boiling-liquid two-phase type or the circulating single-phase type. In the two-phase type, the nitrogen is allowed to pass into the shroud and there to boil at approximately atmospheric pressure. The combined nitrogen gas and remaining liquid is then removed and passed through a reservoir in which the gas is separated from the liquid and allowed to escape. Such systems are usually quite simple since a thermosiphon effect can be utilized to circulate the nitrogen. However, the greatest difficulty in such systems is that any passageway which is slightly warmer than the others will tend to become blocked by gas in the familiar gas-lock fashion sometimes experienced in automobile carburetors. Since there are normally a number of such passageways in parallel in the system, the blocked passageway simply does not pass any liquid and it becomes still warmer, while the nitrogen gas in the passageway prevents the addition of any liquid nitrogen which might cool it down. In consequence, nonuniform temperatures are apt to occur in shrouds utilizing the two-phase systems unless the shrouds are quite small. Also, heat transfer from the two-phase gas-liquid mixture to the shrouds is poorer than in the all-liquid systems, thus resulting in higher shroud temperatures.

Large systems normally make use of so-called "single-phase systems." In these, a liquid-nitrogen subcooler is utilized in which one side of the system is carried under positive pressure. Here, it is possible to cool the liquid nitrogen at perhaps 60 pounds pressure to a sufficiently low temperature that it can be circulated through the system without boiling. When the nitrogen circulates through the shroud, its temperature rises, reaching a value higher than the boiling point of liquid nitrogen at atmospheric pressure. The warm liquid is then returned to the subcooler, where it is once more cooled in a heat exchanger and recirculated. The heat is removed from the heat exchanger by boiling liquid nitrogen outside of the exchanger coils, which are kept at approximately atmospheric pressure, the whole being contained in a vacuum-jacketed dewar. In some cases, the pressure outside the cooling coils in the dewar is actually carried at values below atmospheric through the use of an exhauster which causes the boiling point to be lower than it would be at a pressure of 1 atmosphere, thus cooling the compressed nitrogen still more.

8.10 *Shroud Materials*

Various materials have been considered and used for shrouds in cryogenic systems, but of various possible materials two have emerged as most useful. These are stainless steel and aluminum.

Stainless steel has the advantage of being easily welded into gas-tight systems and of being compatible with passthrough and flange materials so that no difficulty is experienced in connecting the shrouds to the passthroughs. Stainless steel is used frequently for systems employing nitrogen shrouds alone, being generally fabricated of convoluted sheets welded to flat sheets or other convoluted sheets to form the passageways.

The difficulties with the use of stainless steel arise primarily from the relatively low heat conductivity of this material. Passageways for liquid nitrogen must be close together if hot spots are not to develop in between the liquid-cooled passageways. In addition, stainless steel is relatively heavy, which leads either to support problems or to the necessity of using very thin stainless steel for the shrouds, leading to fragility.

Aluminum for shrouds is generally formed by means of the tube-in-sheet process in which two sheets are placed together with a stopoff material placed between them and then bonded by rolling. After rolling, the passageways (where the stop-off material has prevented bonding) are inflated by air pressure to form the interconnecting passages for liquid nitrogen desired. Assembly of such tube-in-sheet arrangements into actual shrouds is relatively simple since aluminum pipes need only be joined at the points where the passageways exit from the sheets.

Because of the high heat conductivity of aluminum and because its light weight permits the use of reasonably thick materials, the passageways can be relatively widely spaced without introducing undue hot-spot effects. Such shrouds have therefore become very common in space simulators and other large vacuum systems.

The use of aluminum shrouds requires some sort of transition device to pass from aluminum piping connecting the shrouds to the stainless steel of the flanges through which the lines enter the chamber. Brazed stainless-steel-to-aluminum transitions have now been developed commercially and seem to have solved this problem in satisfactory fashion.

chapter 9

Vacuum Gauges

9.1 *Introduction*

One of the major problems in all vacuum systems is measuring the pressure within the system after the system has been pumped down. Probably more papers have been written on the problems of vacuum gauging than on all other aspects of vacuum systems combined; and, far from dying out, the interest in the field and in the new inventions made therein continues to proliferate at a rapid rate. As the degree of vacuum achieved becomes better and better, the difficulties of accurately measuring the pressure remaining become even greater and the results increasingly uncertain.

9.2 *Basic Gauge Types*

There are basically four types of gauges which are of importance in connection with vacuum systems. They are as follows:

Absolute-pressure Gauges These are gauges which employ some basic property of nature to measure pressure in an absolute fashion without reference to any other source. They include the mercury manometer, the McLeod gauge, and the Knudsen gauge.

These three types of gauges, while of very great importance, are not sufficiently continuous in operation, sensitive enough, or sufficiently rugged to permit their use on practical operating systems, although they are used in calibrating other types of gauges and will be discussed more fully in Chapter 11.

Thermal-response Gauges Gauges of this type are useful for pressures ranging from a few torr down to 1×10^{-3} or occasionally 1×10^{-4} torr. They are universally used in forelines in vacuum systems, but because of their limited sensitivity are not usually used in the main chambers.

Ionization Gauges These gauges do not actually measure pressure but are used to infer pressure from molecular densities arrived at through ionization phenomena. They are differentiated as (1) the cold-cathode discharge gauges, or Penning magnetron gauges, and (2) the hot-filament gauges, such as the triode gauge and the Bayard-Alpert gauge.

Partial-pressure Gauges These gauges, by various techniques, select molecules of common molecular weight and measure the number of such molecules present, inferring the partial pressure of this particular component of the atmosphere after summing the partial pressure of all species present. Typical examples of this type of gauge are (1) the sector-type magnetic mass spectrometer, (2) the cycloidal-type magnetic mass spectrometer, (3) the quadrupole- and monopole-type mass spectrometers, (4) the time-of-flight mass spectrometer, and (5) the cyclotron-resonance type (omegatron).

9.3 *Thermocouple Gauges*

These small, compact, easily portable, and rugged gauges depend for their principle of operation on the fact that heat flow away from a moderately heated wire will be a function of the number of molecules present to carry away heat by convection, the cooling effect being greater at high pressures than at low. Typically, such a system consists of a heated, blackened wire of tungsten or other metal, thin in cross section, to which is attached a thermocouple or thermocouples whose leads are brought out to an external circuit in such a manner that the temperature of the filament can be read in terms of the pressure of the gases surrounding it at constant electrical input to the filament. Figure 9.1 shows the filament and thermocouple arrangement employed in such a gauge. Generally, the output of the thermocouple is read by a simple millivoltmeter type of instrument which is calibrated in terms of pressure and

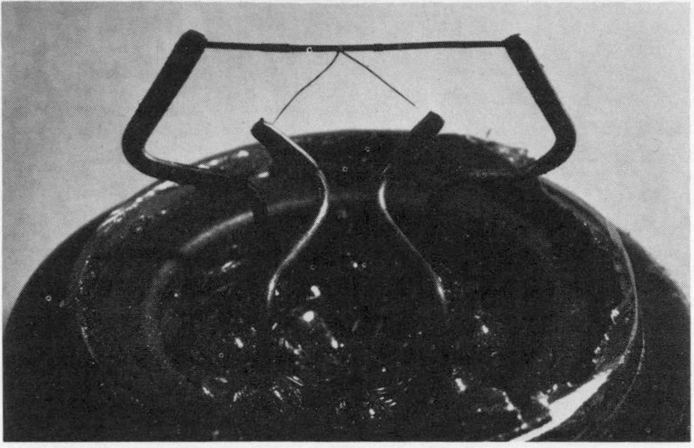

Fig. 9.1 Thermocouple vacuum gauge. (*Bendix Corp., Rochester, N.Y.*)

can be read directly. One such gauge, the Hastings gauge, employs a bridge circuit of the conventional wheatstone type in order to secure greater precision of readout. Such gauges are relatively inexpensive, simple, and rugged in design. While they may be contaminated by material from a chamber or from pump fluids, they can sometimes be cleaned after contamination by a simple washing in acetone followed by general warming for drying. In any event, the sensing units are relatively inexpensive and can be replaced when necessary without great cost.

Most operating systems make use of several such gauges in the roughing lines and in the foreline of the diffusion pump in order to measure the performance of these pumps and make sure that all are operating properly. Frequently, a single readout with several switching positions is employed so that several gauges can be read sequentially.

Such gauges have certain disadvantages. As ordinarily made, their range is relatively narrow, ranging from perhaps 1 torr down to 20 or 25 microns (2 to 2.5×10^{-2} torr). Invariably, the gauge scale will have readings below this point, but careful calibration of a number of these gauges will indicate that below 20 or 25 microns (2 to 2.5×10^{-2} torr) most commercial varieties cease to reflect accurately the pressure, since their sensitivity is not sufficient to indicate the difference in heat conductivity of pressures below this point. The only exception to this rule is the variety of this gauge made by Hastings, which is sufficiently sensitive to read pressures down to 5×10^{-4} torr with fair accuracy. However, as in all else, one gets increased sensitivity on the low side by the loss of ability to read on the high side so that, when using thermocouple gauges to read pressures down to and below 1×10^{-3}

torr, it will be necessary to use a separate and different gauge to read pressures in the high range since the same tube cannot cover both ranges.

Pirani Gauges Pirani gauges are basically similar to the thermocouple gauges described above. Again, a heated filament is used, the temperature of which depends upon the density of the gas surrounding it; but in the case of the Pirani gauge, the temperature of the heated wire is read out by measuring its resistance as one arm of a bridge circuit rather than by the use of an attached thermocouple. Otherwise, the principle of operation is similar to that of the thermocouple gauge. Such gauges exhibit the same limitations as thermocouple gauges.

9.4 *Errors of Thermal-response Gauges*

It should be pointed out that all thermal-response gauges have certain built-in limitations with respect to the manner in which they are used. The heat lost from the heated filament and hence the temperature of the filament, which is used as a pressure indication, is not alone a function of the pressure of the gases within the gauge, since the filament also radiates to the walls of the tube.

Therefore, if the temperature of the housing of the gauge tube is at any temperature other than that for which calibration has been made, an erroneous reading will result. A relatively slight increase or decrease in the ambient temperature surrounding the gauge tube will result in a relatively large change in reading, so that an error of as much as 50 percent of the reading may be introduced by a change of 15°F in ambient temperature.

It is also evident that the thermal conductivity of the residual gas surrounding the hot filament is dependent upon the composition of the gas. Such tubes are normally calibrated with the air as the surrounding medium; and for air or nitrogen the calibrations will be correct, assuming that all other factors are constant. However, if the residual gas is hydrogen or largely hydrogen, the temperature will be lower since the hydrogen will carry heat away more rapidly than air. The gauge will therefore read high. Similar calibration effects will occur for other gases which may exist in the residual atmosphere. The gauge should therefore be calibrated for the residual gas expected.

9.5 *Ionization Gauges*

The kinetic energies acquired by an electron in passing down a difference of V volts corresponds to VE, where E is the charge on the electron. When this energy exceeds a certain critical value corresponding to the ionization potential V_1 of a gas, there is a definite probability, which varies with V and with the nature of the gas, that collision between electrons and molecules will result in the formation of positive ions. A relatively high-velocity electron colliding with a gas molecule drives an electron out of its orbit around the nucleus of the molecule, leaving the molecule positively charged. If the energy of the impinging electron is substantially equal to the ionization potential, the chances are high that the colliding electron will, in turn, be captured, resulting in neutralization of the ion so that no permanent ion results. If, however, the energy of the colliding electron is more than twice the ionizing potential, then the probability is that not only will an electron be ejected from the atomic orbit, but that the impinging electron will also be lost, leaving a permanent ion. The value of the ionization potential V_1 for different monatomic gases varies from 3.88 volts for cesium to 24.58 volts for helium, which has the highest ionizing potential. For oxygen, hydrogen, nitrogen, and other diatomic gases, the values of V_1 range around 15 volts. The probability of creating an ion by collisional processes also varies somewhat with the dimensions of the molecules for the diatomic gases. In any case, in a given gas at low pressure, the number of positive ions produced by the passage of a stream of electrons is directly proportional to the molecular concentration. The linear relationship between ionization and density holds from substantially zero gas pressure to the point at which ion formation is sufficient to alter effectively the current and energy of an electron stream. Since the gas pressure at a constant temperature is directly proportional to its density, the positive ion current produced by a steady electron current may be used as an indicator of the pressure.

A triode tube, as shown in Fig. 9.2, provides a simple and convenient means for producing a constant electron stream and separating from

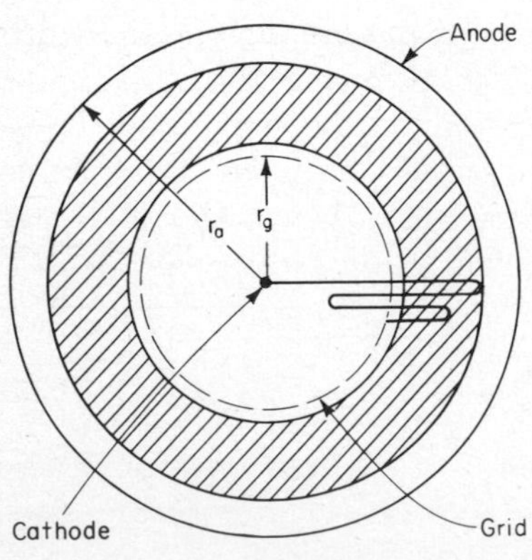

Fig. 9.2 Hot-cathode triode ionization gauge showing a typical electron trajectory. Shaded area indicates useful ionizing region. Voltages: grid-cathode 200, anode-cathode 20.

it the resulting ion current. With the electrode potential at the proper value, all the electrons emitted from the thermionic cathode must reach the grid. Many electrons, however, do not directly collide with the grid wires on the first pass, but oscillate about the grid as indicated in Fig. 9.2, forming an electron cloud distributed throughout the area shown shaded in the diagram. All the positive ions produced between the grid and anode by this electron cloud travel down the potential gradient to the anode. For constant electron current and fixed electrode potential the number of ions produced between the grid and anode (and therefore the anode current) is directly proportional to the gas pressure. The complex relationship between efficiency of ionization and the electron energy means it is not practical to calculate the sensitivity for the simple triode gauge, even after making simplifying assumptions about the electron trajectory. It is more convenient, therefore, to calibrate such gauges against some standard instrument, for example a McLeod gauge. In commercial gauges, sensitivities at their fixed recommended emission values are frequently given in terms of amperes per torr, which is convenient for use in actually reading the gauge.

9.6 *Relative Sensitivity for Different Gases*

The measurement of the probability of ionization shows that the gauge sensitivity depends upon gas composition. Further, the relative sensitivity *R*, the ratio of sensitivity for a given gas to the sensitivity for nitrogen, is a function of electron energy. This function is not a constant for all gauges but varies depending upon the geometry of the gauge and should therefore be determined for the gauge being tested or obtained from the manufacturer rather than be used from general data applying to some different gauge. Approximate values of *R* as determined for a number of gauges by a number of different investigators are given in Table 9.1. The last column of this table provides approximate values for Bayard-Alpert gauges, as most commonly used by various laboratories.

It should be reemphasized that these values are approximations which will vary with the gauge being used and in any case are imperfectly known. Where precision in the values is essential, calibration with pure gases should be carried out.

It should be pointed out here that, since in most vacuum systems the actual composition of the gases is unknown, this correction cannot be made; therefore, there usually is doubt as to exactly what the pressure is as read from an ionization gauge. Where mass spectrometers are

TABLE 9.1 Approximate Relative Sensitivity of Gauges

Gas	Observer					B-A gauge suggested usage
	Dushman and Young	Reynolds	Langmuir	Redde-ford	Metson	
Nitrogen	1.00	1.00	1.00	1.00	1.00	1.00
Oxygen	....	....	....	0.14	0.79	1.86
Hydrogen	....	0.46	0.38	0.38	0.52	0.38
Carbon monoxide	....	....	....	....	1.05	1.05
Carbon dioxide	....	....	....	....	1.36	1.36
Water vapor	....	....	....	....	2.0	2.0
Dry air	....	....	....	0.81	....	0.90
Mercury	3.3	3.44	2.1	....	....	3.4
Helium	0.18	0.16	0.15	0.25	....	0.16
Neon	0.35	0.24	0.23	....	....	0.24
Argon	1.4	1.19	1.11	1.06	....	1.2
Hydrocarbons: oil fragments and miscellaneous contaminants	....	....	....	....	....	5.0

employed to determine the actual residual gas composition, these corrections can be profitably used. Where such an instrument is not available, it is customary to report pressures in terms of "equivalent nitrogen pressures," meaning that the pressures reported are those which would be correct if the residual gas were entirely nitrogen and that no correction has been applied for gas composition.

9.7 *Bayard-Alpert Gauges*

At first sight, there appears to be no lower limit to the pressure range of the gauge described previously (the triode gauge), provided that a sufficiently sensitive instrument is available to measure the ionization current. In practice, however, a definite lower limit is set for these gauges due to the x-ray effect.

The x-ray effect is caused by electrons from the filament striking the grid with sufficient energy to excite the emission of soft x-rays. When these reach the collector (a circular cylinder in the triode gauge) they cause the photoemission of electrons, which are electrically indistinguishable from the collection of positive ions.

Arbitrarily, the photoemission limit or x-ray limit in a gauge has been chosen as that pressure at which 10 percent of the anode current is caused by this x-ray or photoemission effect. When this occurs, the

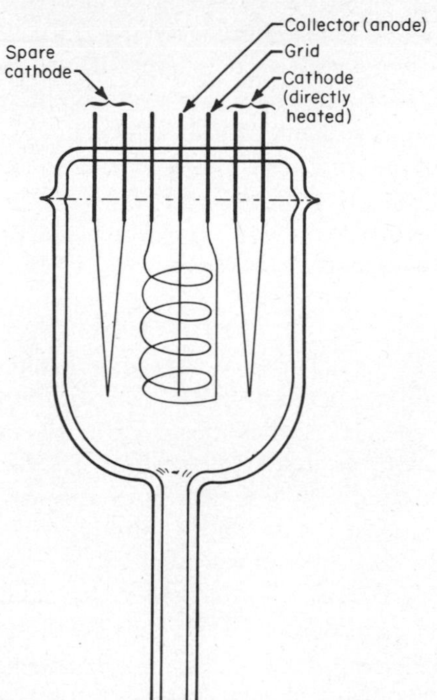

Fig. 9.3 Bayard-Alpert ionization-gauge tube with inverted structure and extended range.

gauge is said to have reached its x-ray limit. For triode gauges, as normally constructed, similar to the sketch shown in Fig. 9.2, this limit is approximately 2×10^{-7} torr. If the system pressure falls below this point, the gauge will normally not follow the pressure down but will continue to read a pressure in the neighborhood of 1×10^{-7} torr regardless of system pressure below this value.

Bayard and Alpert devised a different form of gauge which achieved a great improvement with respect to the x-ray effect. In their gauge, the filament is placed outside the grid instead of centrally within it. A thin wire, centrally located within the grid, serves as the anode. The ionization region in this gauge, as shown in Fig. 9.3, is the region within the cylindrical grid in which the electrons oscillate, creating ions which are then collected by the thin-wire anode. With this gauge, sensitivity is approximately the same as that of a standard triode gauge, but because of the very small surface area of the anode the photocurrent is reduced by some two or more orders of magnitude. With an accelerating voltage between the grid and filament of +200 volts and with an anode-filament voltage of —20 volts, a rather high potential gradient is formed close to the anode so that the electrons travel almost to this collector with approximately constant energy at about 200 electron volts. That is about the level required for most efficient ion production. This is a distinct advantage over the conventional gauge, where the potential, and therefore electron energy, falls approximately linearly between the grid and the anode so that efficient ion production cannot take place throughout the whole volume.

Such Bayard-Alpert gauges have become the prime method of measuring pressures in the vacuum systems operating between 1×10^{-4} and 5×10^{-10} torr. Their response to pressure over this range is essentially linear, except at the very highest portion of the 1×10^{-4} torr range, and the ability to clean up the gauge by electron bombardment techniques from the filament or by resistance heating of the grid structure has made them suitable for most commercial applications. Various commercial examples are made with heavier or lighter electrode wires and other parts, so that they will stand more or less vibration and mishandling with consequent changes in the x-ray limits introduced by these effects; thus, various examples may have an x-ray limit ranging from approximately 3×10^{-9} to 1×10^{-10} torr or even lower. In addition, more than one filament can be placed around the outside of the grid cage so that the burnout of one filament does not necessarily require discarding the gauge, since the second filament can be lighted and the work carried on without releasing the vacuum.

9.8 *The Nottingham Gauge*

One difficulty encountered in the Bayard-Alpert-type gauge is the fact that a space charge is built up on the glass envelope of the gauge, and this space charge is the cause of instability at low pressure.

Nottingham and Alpert have modified the Bayard-Alpert gauge to improve the sensitivity. The cylindrical grid is closed at the top and bottom, and a second grid, acting as a screen grid, is installed around all of the electrodes or deposited as a conductive coating on the glass-tube walls. The purpose of closing the cylindrical grid is to prevent ions from escaping to the negatively charged glass wall. At low pressures, the ions may oscillate about the collector wire many times before being collected. Since the ions are likely to have some component of their velocity parallel to the collector wire, they may escape through the ends of the grid before being collected unless the latter is closed. A screen grid or deposit on the walls shields the grid from wall charges, being operated at a negative potential, thus increasing the average path length of the electrons and giving the gauge a higher sensitivity. At the same time, the screen grid also helps to prevent the buildup of space charge on the envelope, which would then have the effect of

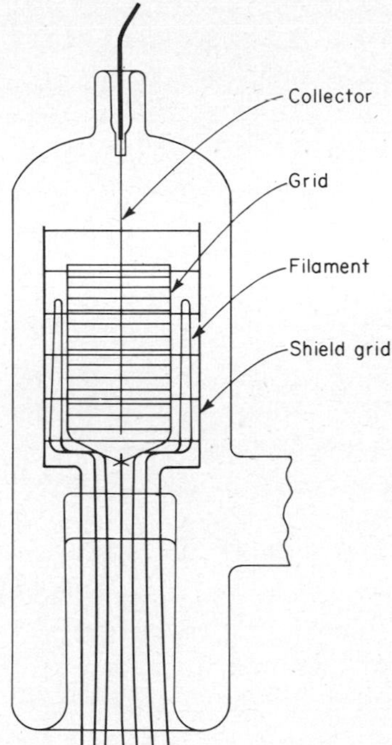

Fig. 9.4 Nottingham ionization gauge.

lowering the x-ray limit of the tube. Figure 9.4 shows one version of the Nottingham arrangement of the Bayard-Alpert gauge.

9.9 *Cold-cathode Ionization Gauges*

The practicability of using a cold-cathode glow discharge to measure gas pressure in regions of 1×10^{-3} torr was first realized by Penning. He used a strong magnetic field to increase the electron path length from cathode to anode and so raised the ionization to a measurable level. In 1937, Penning described a practical gauge in which a discharge took place between an anode made in the form of a loop and two zirconium disks connected together electrically, one on each side of the ring, acting as cathodes. A permanent magnet is placed with the field parallel to the axis of symmetry. In this gauge, electrons originating from either of the cathodes are prevented from going directly to the anode by the magnetic field. They travel, instead, in helical paths back and forth in the potential trough between the cathodes before eventually drifting to the anode. Due to the long electron path, there is a high probability of an ionization collision with a gas molecule,

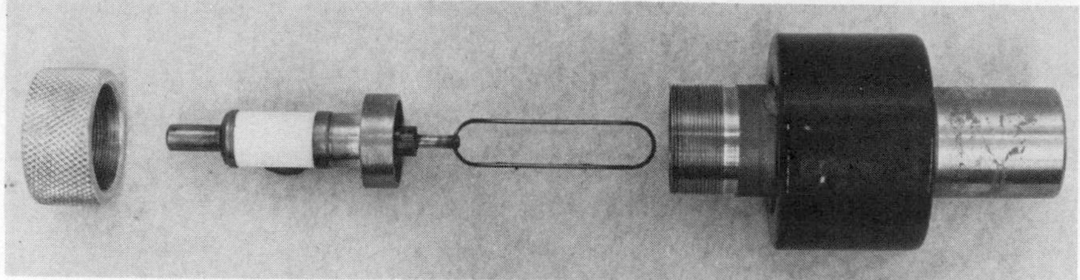

Fig. 9.5 Penning cold-cathode ionization gauge.

even at 1×10^{-5} torr. The positive ions, being virtually unaffected by the magnetic field because of their large mass, travel directly to the cathode. Secondary electrons released from the cathode by the positive-ion bombardment serve to build up and maintain the discharge. Thus, with the help of a magnetic field, the glow-type discharge is maintained, even with the electron mean free path of the gas many times greater than the distance between the anode and the cathode. Figure 9.5 shows a disassembled view of a cold-cathode gauge.

The use of gauges of this type is generally limited to pressures of 1×10^{-5} torr and above, since the sensitivity of the gauge deteriorates to a point where it cannot be used at lower pressures. However, the cold-cathode gauge has certain very definite advantages which render it suitable for certain types of work.

It is a very simple structure which can easily be disassembled and cleaned should contamination occur, which is certainly not the case with the Bayard-Alpert gauge. It is, moreover, immune to damage should the pressure suddenly come up to atmospheric and can be restarted without difficulty when pressure is again brought down to the vacuum region. It is therefore useful as a gauge to cross over the junction between the highest pressure readable with hot-filament gauges and that measurable by Pirani or thermocouple gauges and has had considerable application for this type of work.

9.10 *Cold-cathode Inverted Magnetron Gauge*

The Penning type of cold-cathode discharge gauge would appear to have certain advantages over the hot-cathode ionization gauge for measurements in the ultrahigh-vacuum region. There is no limitation of pressure measurement due to x-ray photoemission since the number of electrons to strike the anode and produce x-rays decreases with the pressure and remains constant as a percentage of pressure. However, there are certain inherent disadvantages in the operation of the Penning gauge at pressures below 1×10^{-5} torr. In most gauges, the discharge fails to strike below these pressures. The application of higher voltage to start and maintain a discharge leads to field emission from the ion collector which cannot be distinguished from the ion current by the external measuring circuit. This field-emission current establishes a lower pressure limit for the operation of the normal Penning gauge, analogous to the x-ray limit for the Bayard-Alpert gauge. To circumvent these difficulties, Hobson and Redhead designed an ionization gauge employing a cold-cathode discharge at crossed electric and magnetic fields. Their gauge, shown in Fig. 9.6, has the structure of an inverted magnetron with an auxiliary cathode. This geometry provides efficient electron trapping in the discharge region, and the auxiliary cathode provides the initial field emission for starting and allows the positive ions to be measured independently of the field-emission current. The auxiliary cathode also acts as an electrostatic shield for the ion collector; and the two short tubular shields, which project 2 millimeters into the ion collector from the auxiliary cathode, protect the end plates of the ion collector from the high electric fields and provide the field emission which initiates the discharge. The gauge operates in an applied magnetic field of 2,060 oersteds and a potential of 6 kilovolts on the anode. Under these conditions, the relationship between ion-collected current and pressure is given by $I_p = cp^n$, where p is the pressure, c is a constant, and n lies between 1.10 and 1.40 for various gauges.

With this gauge, the time lag between the application of anode voltage to the gauge and the initiation of stable discharge becomes appreciable and variable at pressures below 1×10^{-8} torr. At pressures in the 1×10^{-10}-torr region and below, the gauge generally cannot be started without some auxiliary source such as an ultraviolet lamp or Tesla coil discharge. Hobson and Redhead have shown the inverted magnetron gauge to be useful in the pressure range of 1×10^{-3} to 1×10^{-12} torr. The lower limit has not been determined with certainty and almost surely extends downward to 1×10^{-14} torr if sufficiently sensitive readout equipment is used.

Two gauges similar to Redhead's but somewhat different in detailed design have appeared. One of these, the Kreisman gauge, has an essentially similar structure but is housed within a metal encapsulation of quite small size and very rugged construction. Because of its small size, the magnet required may be supported by the gauge itself rather than by an external stand which is necessary with the glass-enclosed Redhead structure. Furthermore, bakeout of the gauge, which is required prior to use at pressures below 1×10^{-9} torr, may be carried out much more rapidly and easily with the metal-encapsulated gauge than with the glass-tubulated gauge of the original Redhead design.

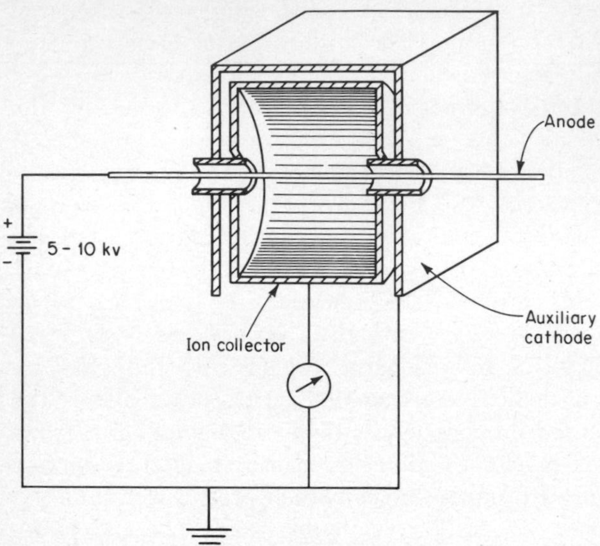

Fig. 9.6 Inverted magnetron gauge (Hobson and Redhead type).

Young of General Electric has devised another cold-cathode gauge which employs cylindrical electrodes and a circular ring-type Alnico V magnet. Figure 9.7 shows a schematic of this gauge, known as the trigger gauge. With this gauge, starting is assured at all pressures through the addition of a small tungsten filament mounted so that, when energized, the stream of electrons can enter the magnetron gauge through a properly aligned hole of very small dimension. The filament is energized only momentarily by a push-button arrangement, so that starting is initiated and the filament is then turned off. With this gauge the tubulation is full diameter into a metal flange so that impedance problems are not serious. The sensitivity is approximately 2.5 amperes

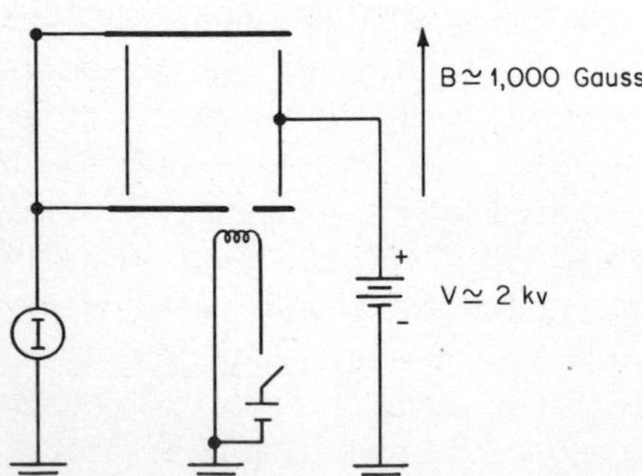

Fig. 9.7 Triggered Penning cold-cathode gauge. (*Vacuum Products Division, General Electric Co.*)

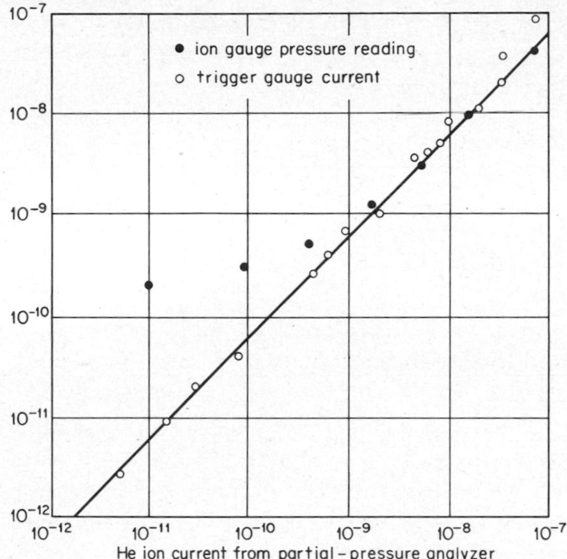

Fig. 9.8 Calibration curve on triggered Penning gauge. (*Vacuum Products Division, General Electric Co.*)

per torr at 2,000 volts potential, and satisfactory operation below 1×10^{-13} torr has been readily attained. A calibration curve for this gauge for helium, obtained by comparison with a sensitive partial-pressure analyzer set on the helium line, is shown in Fig. 9.8. The gauge seems to be usable at pressures below 1×10^{-13} torr.

Further study by Paul Bryant and his associates at Midwest Research Institute, as well as by Young and others at G.E., has shown that this gauge as well as the Redhead gauge is subject to numerous discontinuities in gauge constant representing changes in the discharge mode. This is especially true at pressures below 1×10^{-9} torr. These mode changes have the effect of changing the exponent in the gauge equation $I_P = CP^N$ from about 1.1 to 1.4 or 1.6. Mode changes take place seemingly erratically during pumpdown, causing gauge-current oscillations with periods ranging from 30 seconds to several minutes. All examples of the gauge do not perform identically, nor does the same gauge duplicate its performance each time it is cycled.

It is the opinion of the writer that many of these erratic performances are due to the presence of minute amounts of organic contaminants within the gauge structure, where they can be activated by the discharge. No internal means of bakeout is available with either the trigger or the Redhead tube to clean up the gauge. However, an external bakeout oven or heater can be used after removal of the magnet to clean up the tube during early portions of the pumpdown. The pressure should be at the 1×10^{-5}-torr level or lower when the bakeout is carried out, or else more contamination will take place as the gauge cools. Bakeout during the last portions of chamber bakeout is best. The gauges seem to operate with reasonable consistency after such treatment, even down to the bottom of the 1×10^{-11}-torr range.

9.11 *The Extractor Gauge*

A new type of hot-filament ion gauge developed by Helmer and Hayward, and offered by Varian, makes use of the extractor principle described by Scheumann. In this gauge the ions are generated by the usual mechanism similar to that used in the nude Bayard-Alpert gauge. However, instead of collecting these by an internal collector, they are drawn out by an electrostatic field, bent through an angle of 90 degrees, then after passing through a slit are collected by a flat-plate collector. Due to the bent-beam design, the x-ray current is reduced by several orders of magnitude from that observed in the Bayard-Alpert gauge of conventional construction. Accurate calibration techniques are not available to assess the accuracy of this gauge, but it appears to be useful to at least 1×10^{-13} torr and possibly as low as 1×10^{-14} torr. Further improvements with gauges built by this method seem able, at present, to keep up with the improvement in system pressure capabilities, at least for another year or two.

9.12 *Gauge Readout Devices*

Regardless of the type of gauge sensor used, the actual accuracy in use will depend upon the sensitivity and linearity of the readout device used. Commercial readouts as furnished with the gauges are of reasonable accuracy but have certain weaknesses inherent in their designs due to the necessity for low price. For very accurate work, these readout devices should be checked against standard electrical instrumentation, calibrated against National Bureau of Standards instruments, or eliminated and special precision-calibrated readout devices used. Critical points to be checked are (1) uniformity and stability of the power supply, providing the accelerating voltages, (2) linearity and stability of the electrometer amplifiers used in providing the readout indication, (3) accuracy and linearity of the readout milliammeter, and (4) accuracy and stability of the decade-adjustment resistor used.

In most commercial gauges, the various decade switches are arranged so that they do not overlap, meaning that readings must be carried down to the 10 percent limit before the next more sensitive scale can be used. Unfortunately, accuracies which may be quite satisfactory as a percentage of full scale become rather unsatisfactory at the 10 percent read point, so that limited accuracies are usually obtained in commercial instruments. Precision micro-microammeters generally provide a one-third overlap in the decade adjustment to avoid this type of difficulty.

chapter 10

Vacuum Gas Analyzers

10.1 *Introduction*

In order to measure the composition of gases remaining within a vacuum system at low pressures, some form of partial-pressure measuring device may be used which ionizes some portion of the gases present, separates them into groups having the same mass to charge ratio m/e, and counts the population of each group. Such devices are often called "residual gas analyzers," "partial-pressure gauges," "mass filters," or simply "mass spectrometers." An official name, *vacuum gas analyzer*, has been adopted by the American Vacuum Society and the American Society for Testing and Materials. We shall therefore hereafter refer to any such device as a "VGA."

10.2 *Sector-type VGAs*

Basically, a VGA is a simplified mass spectrometer with the omission of the usual vacuum system and the specimen admission system. In its oldest and simplest form, the instrument has only three parts: an ionizing section where ions are produced by a beam of electrons and from which the ions are ejected by a variable repeller voltage; a sector in which the ions are constrained to travel in an arc determined by their mass and a magnetic field applied at right angles to their direction of travel; and a slit and collector for counting the ions. In most instruments of the sector type, sweeping (scanning from one end to the other of the atomic spectrum) is accomplished by changes in the accelerating voltage, which has the result of bringing various-weight molecules to the slit leading to the collector. Usually, the magnetic field is held constant, and the mass number varied from 2 to about 60 by changes in voltage. For heavier masses, either a stronger permanent magnet or an electromagnet is used. The latter option is somewhat more effective, since the field and the voltage can both be utilized in sweeping. However, a more elaborate power supply is required, which considerably increases the expense. As a result the electromagnet is little used on VGAs.

In order to increase the sensitivity of an analyzer, an electron multiplier may be provided at the collector location, which increases the sensitivity by several orders of magnitude at the expense of some degree of accuracy and repeatability, since fully stable and constant multipliers are yet to be made.

10.3 *Cycloidal-type VGAs*

The cycloidal type of mass spectrometer is somewhat different in principle from the plain sector type, but again uses voltage sweep to control the motion of ions in a fixed magnetic field. This instrument is widely used, and thousands of cracking patterns are available for it, so that it is a very useful instrument within its range (see Sec. 10.7 for more on cracking patterns).

10.4 *Omegatron-type VGAs*

The omegatron is an inexpensive type of VGA which is widely used in college and university laboratories, where the source of high-frequency power required is often available and does not have to be purchased for the VGA alone. This instrument produces ions in a conventional manner near the center of a small box. Selected masses are accelerated by a radio-frequency field and constrained by a fixed magnetic field in such a fashion as to excite them into their characteristic cyclotron resonance. Under these conditions the ions of interest spiral outward until they reach a collector on the outer wall of the box. Ions whose characteristic resonance frequency does *not* match that of the applied electric field spiral out a short distance, then spiral in again, never reaching the collector. The chief difficulty with the omegatron is that the box in which the action takes place is almost completely closed and has an extremely low conductance to the chamber being monitored. Under these circumstances, the resemblance between the omegatron gases and those in the chamber may be rather remote, especially in chambers where high pumping speed is taking large amounts of gas from the test item through and out of the chamber so that conditions in the chamber never reach equilibrium. However, it may be completely inserted within the chamber and is so used in many experimental setups, especially in small vacuum evaporators. Unless a high temperature bakeout is carried out—often as high as 800°C—a variety of spurious peaks are often present, representing doubly and triply ionized molecules.

10.5 *Time-of-flight VGAs*

The time-of-flight mass spectrometer is the most expensive of the instruments used for VGAs. It makes a direct measurement of ion speed as the ion passes down a flight tube and infers the ion m/e from its flight time. The ionizing electron beam is pulsed at a high rate, and the arrival of each group of ions of the same m/e ratio is observed by a high-gain electron multiplier. With this instrument a very rapid scan rate is possible, making cathode-ray-tube display of very rapidly varying phenomena possible. Bakeout is a problem, since all parts, including the long flight tube, must be fully degassed.

10.6 *Quadrupole and Monopole VGAs*

A recent development in the field is the quadrupole, or the related monopole, type of mass spectrometer. These instruments make use of dc and ac voltages applied to oppositely placed pairs of accurately spaced stainless steel rods. Ions are produced conventionally and injected into the quadrupole field at one end of the structure. By properly controlling the dc and ac voltages and ac frequency, one, and only one, type of ion will pass in a stable manner down the axis of the rods, all other m/e ratios being unstable and rejected to one side or another. Electron multipliers are used on most such units to increase the sensitivity.

The problems in the use of VGAs in vacuum systems are many, and difficult to deal with. We shall list a few of them with our comments.

10.7 *Problems in Using VGAs*

1. *Bakeout Problems* Mass spectrometers or partial-pressure gauges, like vacuum gauges in general, do not read the gases within the system but those present in the ionizing chamber, plus those produced within the system which are able to reach the collector. Thus careful and lengthy bakeout is almost always required for accurate work. Molecules originating within the instrument itself—carbon, silicon, potassium, sodium, and a variety of hydrocarbon molecules or fragments—must be eliminated by design and baking before the reading can have much validity. The variety of molecules reported in some papers of recent vintage indicates that this objective is not always reached in practice.

2. *Access of Gases from the Chamber to the VGA* If the instrument is designed so that long, small-diameter tubing must be employed to connect the VGA ionizing chamber and the chamber being analyzed, any resemblance between the mixture of gases in the chamber and that in the VGA may be purely coincidental. A nude ion source, located if possible within the chamber, is a necessity for accurate work.

3. *Amplifier Stability* The electron multiplier, if used, and the amplifier in any case must be highly stable and linear. The results obtained should be recorded so as to permit later peak identification and measurement. Generally, an X/Y recorder plotting accelerating voltage against peak height is most satisfactory. Where very rapidly changing phenomena are to be recorded, high-speed photography of an oscilloscope may be the only method possible.

4. *Analyzing the Readings* A mere reading of peak heights obtained in a scan by a VGA, no matter how accurate, does not constitute an analysis. Each type of molecule entering the instrument will, in general, give rise to more than one peak. Thus water, H_2O, will give rise to minor peaks at a m/e of 16 (O^+) and at a m/e of 17 (OH^+) as well as at a m/e of 18 (H_2O^+). In addition a minor peak will be found at a m/e of 2 (H_2^+). The relative peak heights for a given entering group of molecules are known as the "cracking pattern" for that molecule. Frequently, parts of these cracking patterns overlap those produced by another molecule and are thus not separable. A familiar example of this effect is that found at mass 28. A peak at this point may be due to either N_2^+ or to CO^+, and the difference between the two is so slight as to be indistinguishable in a normal VGA. However, if the cracking pattern for nitrogen is available, the ratio between N^+ and N_2^+ will be known. By multiplying the reading at $m/e = 14$ by an appropriate number, that portion of mass 28 due to nitrogen can be determined. The remainder will be that due to CO^+. By working through such cracking patterns for all molecules suspected to be present, the peak height due to each molecule may be determined. From this, the partial pressure due to each species may be obtained. Alternatively, results may be stated as a percentage of some constant gas, usually nitrogen or helium.

The point of importance is that while the indicated readings may be useful for leak detection, troubleshooting, etc. they are *not* analyses of the residual gases present until cracking pattern corrections have been made. The time involved for a computation by hand is long; however, computer programs have been developed to speed up the job. If an "on-line" computer can be used, real-time results, limited only by printer speeds, can be obtained.

10.8 *Specifying VGA Characteristics*

At present, there is no standard way of specifying VGA characteristics. Each vendor has used his own definitions, so that direct comparison is difficult.

1. *Resolving Power* This term is used by various vendors to mean:

a. The ability to separate masses differing by one mass unit widely enough so the valley between adjacent peaks is not more than 1 percent of the height of the larger peak.

b. The ability to separate peaks one mass unit apart widely enough so that the larger contributes not more than a fixed amount—usually 30 percent—to the smaller peak.

c. Measurement of the absolute height of two adjacent peaks, even though the valley between may be as much as 90 percent of the peak heights.

Such a consideration is meaningful only if the mass range at which the reading is taken is stated and if the required total pressure in the system being measured is stated. Sector-type instruments do not provide uniform spacing of peaks over the entire range but crowd peaks closer and closer as the mass grows greater. Quadrupole-type instruments separate masses uniformly but tend to have a varying sensitivity at different mass numbers.

2. *Unit Resolution* This term *usually* means that the instrument can separate peaks one unit apart according to one of the above definitions. Usually it is stated as "unit resolution to mass 50," meaning that masses differing by one mass unit can be separated, from mass 2 to mass 50 inclusive. To be complete the required system pressure should be stated.

3. ***Sensitivity*** This is generally stated in terms of minimum detectible partial pressure, as "minimum partial pressure 1×10^{-10} torr." It is meaningful only if the required system pressure to reach this sensitivity is given, since at higher pressures the system noise may make reaching such a limit impossible. Amplifier noise and electron-multiplier noise (if used) must of course be lower than the minimum pressure to be detected.

Note that the minimum partial pressure must be at least two orders of magnitude lower than the total pressure at which the gauge is to be used if constituents contributing 1 percent to the total pressure are to be detected and measured. An instrument with the sensitivity mentioned above (1×10^{-10} torr) should therefore be used at system pressures no lower than 1×10^{-8} torr.

Based on instruments presently available, system pressures down to 1×10^{-10} torr are about the lowest that can be reliably measured. However, much work is being done in improving VGAs, and the reader is urged to consult the manufacturers for later data, since all available texts are apt to be out of date by the time they appear.

For more information on the subject of VGAs the reader may refer to the references given in the Bibliography in the back of this book, especially to Van Atta, Guthrie, and Roberts and Vanderslice. However, the field is advancing very rapidly, so that the best source of information is the vendors of such equipment.

chapter 11

Vacuum Gauge Calibration

11.1 *Introduction*

As was mentioned in the preceding chapter, the phenomena which are used as pressure indications in vacuum systems, namely thermoconduction and ionization effects, are not sufficiently well defined to permit direct calibration of the gauges by mathematical procedures based on first principles. It is therefore necessary to calibrate these gauges by means of some test procedure by which their reading may be compared with a known pressure to which they are exposed and a calibration scale thus derived. Further, it is highly desirable that the calibration be used over the linear portion of any such calibration curve with the gauge not used at the portions of the curve where the slope is nonlinear, since in these regions very small differences in gauge response produce large differences in apparent pressure reading.

It is further necessary to calibrate the readout device separately from the gauge tube or sensor, since it is usually necessary to use the same readout device with one or more tubes, either to reduce the cost of the gauging operations or to make use of a secondary installed sensor when the primary sensor burns out or becomes inoperable due to contamination or other causes. Where gauge tubes and readouts are calibrated together, the pair must remain together in use; and a change in the gauge tube will require recalibration. In the following, we shall discuss the calibration of gauge tubes separately from the readout or electronic devices which are used to determine the output, and hence infer the pressure.

Gauge-sensor manufacturers conventionally test and calibrate typical tubes from each supposedly identical lot, but do not (unless special provisions are made) calibrate each individual tube or sensor. Variations, therefore, exist in the basic sensitivity of individual sensors as purchased, even though the type has been calibrated and all tubes of the lot are supposedly similar. For these reasons, accurate measurement of vacuums requires the provision in the laboratory of a calibration device or sending of sensors to be used to a calibration source and return of the tubes for later use.

11.2 *Calibration by Comparison*

One of the obvious ways by which to calibrate a gauge tube or sensor is to attach it to a system to which an absolute standard gauge has been fastened and simply compare the reading of the standard against the tube being calibrated. Two types of absolute standard devices are available, each of which has severe limitations and difficulties associated with its use. These were mentioned, but not described, in Sec. 9.2, since they are used, in general, only for calibration purposes.

11.3 *The Knudsen Gauge*

The Knudsen gauge is a very delicate gauge and, while it gives absolute values, it is so extremely sensitive that it can be used only in the most favorable controlled-temperature, controlled-vibration laboratory conditions. Even the slightest vibration in the room in which the calibration is being carried out will cause fluctuations in the arm and invalidate the readings. Temperature changes in the room will, of course, also cause difficulties. For this reason, the gauge is not useful for ordinary calibration purposes unless an underground, carefully temperature-controlled vault free of vibrations can be used, which is not the case in most laboratories. It has, therefore, passed completely out of use for calibration purposes.

11.4 *The McLeod Gauge*

This gauge is probably one of the oldest types in use for measuring pressures in vacuum systems. It was developed in the 1890s by McLeod and has since come into widespread use. It is made in several versions, some of which are used at pressures from a few torr down to 1 micron (1×10^{-3} torr) and others which are used from 1 micron (1×10^{-3} torr) down to approximately 1×10^{-5} torr. Such gauges conventionally have scales permitting the reading of pressures in the 1×10^{-6}-torr range and occasionally as low as 1×10^{-7} torr. However, the inaccuracies of the process, which will be described in more detail later, generally prohibit its use at pressures below 1×10^{-5} torr.

One of the most precise of the McLeod gauges is that due to Paul Rosenberg and commercially marketed in the United States by the Bendix Corporation. This arrangement is as shown in Fig. 11.1.

The basic principle of this gauge is very simple. It consists of a large mercury reservoir, a sampling volume, and a method of raising the level of the mercury in such a fashion as to trap the sampling volume, completely isolating it from the vacuum system. The volume is then compressed upwards into a small capillary tube while another arm of the system carries the column of mercury upwards to a feduciary mark on the sidearm capillary, at which point the height of the mercury column within the closed capillary is compared with the height of the

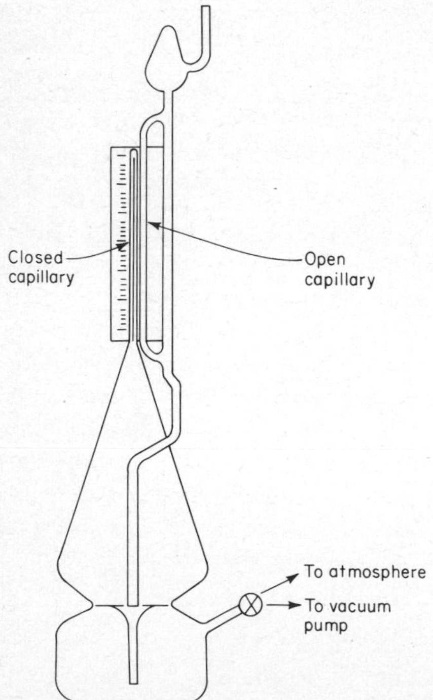

Fig. 11.1 McLeod gauge.

mercury in the open capillary. The formulas necessary for the use of this gauge are derived as follows:

Derivation of Formula for the McLeod Gauge Defining the following terms,

P_1 = unknown pressure
V_1 = volume of gauge from top of capillary to reservoir neck
V_2 = final volume of capillary
P_2 = final pressure in capillary

From the kinetic theory of gases we have, at constant temperature, the following:

$$P_1V_1 = P_2V_2 \tag{11.1}$$

or

$$P_1 = \frac{P_2V_2}{V_1} \tag{11.2}$$

Now

$$V_2 = \frac{\pi D^2 h}{4} \tag{11.3}$$

where D = diameter of closed capillary tube
h = distance between top of mercury column and top of closed capillary tube

and

$$P_2 = h^1 d \tag{11.4}$$

where h^1 is the difference in height of the mercury column between the open capillary tube and the closed capillary tube. If the mercury in the open tube is at the same level as the top of the closed tube, then

$$h^1 = h \tag{11.5}$$

and

$$P_1 = \frac{h^2 d\pi D^2}{4V_1} \tag{11.6}$$

Since the only facts that have to be known to determine the pressure are measurable and are traceable to the National Bureau of Standards by dimensional measurement, the gauge can be used as an absolute standard. Also, the readings are not affected by different gases, provided these gases are not condensible at the pressure of the system. However, since the gas in the McLeod gauge is compressed from a low pressure to a relatively high pressure, any condensible vapors will be condensed at the high pressure in the closed capillary, causing erroneous readings. These vapors consist of such materials as water vapor, pump-oil vapors, etc. Temperature also has a considerable effect on the McLeod gauge reading, since any change in temperature will cause a change in the density of the mercury, which will consequently affect the height read. Temperature changes will also cause changes in the dimensions of the glass parts of the gauge, which will affect readings in an obvious way.

11.5 *Errors of the McLeod Gauge*

It is sometimes assumed that, since the McLeod gauge is an absolute standard, it should be used for all gauge calibration within the pressure range for which it is useful. It is, therefore, important to set forth in some detail the various types of errors to which this gauge is prone and to indicate some of the problems.

1. *Sticking Effects* The mercury rising through the very small closed capillary often does not rise in a slow and uniform manner, but instead is retarded irregularly, giving rise to erroneous readings. The most common cause of this sticking effect is due to the presence of contaminating substances (particularly oils and greases) on the walls of the capillary tube or bits of foreign matter (mercuric oxide, etc.) in the mercury, which can cause plugging or partial plugging of the tube. The mercury used can be of the triple-distilled type and thoroughly cleaned and filtered as it is placed in the tube. However, even when it is so cleaned, some contamination still occurs with time, resulting in the floating of various particulate matter on the surface of the mercury pool, where it can be seen when the mercury is lowered. Some of this material will invariably get into the capillary, causing trouble. As a result, it is necessary to remove the mercury at reasonably frequent intervals and replace it with fresh material, redistilling the removed mercury for reuse.

The contaminating matter on the walls of the capillary can only be removed by prolonged baking, which will probably require raising the McLeod gauge to a temperature of not less than 250°C, meanwhile maintaining the mercury in the reservoir at room temperature by some form of insulation and cooling. Under these conditions, the upper portion of the McLeod gauge expands due to heating, while the bottom portion of the glassware remains at room temperature, and therefore unexpanded. The stresses set up in the gauge by such baking operations are quite high and breakage of the gauge during bakeout is a constant hazard. However, there exists no other way of removing the capillary contamination from a completely assembled gauge.

2. *Static Charges* In addition to sticking due to the problems mentioned, it has been found that the repeated motion of the mercury up and down the capillary tube tends to induce a static charge on the glass of the tube, which then acts to inhibit the ability of the mercury to move in a free manner. Since the glass is an insulator, the removal of the static charge is indeed difficult, and in repeated operations this form of aberration turns out to be one of the major difficulties.

3. *Inaccuracy of the Feducial Mark* It was stated earlier that the mercury in the comparison capillary, or in the main sidearm tube, was brought up to a feducial mark. This mark is so placed by the manufacturer that it represents the top of the closed capillary. Unfortunately, this is an ambiguous term, since the capillary is made by taking a straight length of capillary tubing whose bore is carefully measured by the progress of a mercury drop down the tube while it is observed by a measuring microscope, after which the end of the tube is closed. In order to close the end, heat must be applied to soften the glass, and the forces resulting from this heating and closing of the end tend to change the diameter of the capillary slightly at this point. It is therefore very difficult indeed to get an unambiguous point which represents the actual top of the capillary by any observational means. Normally, this value is obtained by solving of simultaneous equations involved in several readings of the gauge against a known pressure which can yield a value for the "effective top" which is then marked at the feducial mark. It is, of course, quite possible to operate such gauges without the use of a feducial mark, since the equations given previously can be expanded to include readings from both tubes in such a fashion that no set value has to be used. However, by so doing, it becomes impossible to use a fixed readout scale, which is a convenience to be desired for a McLeod that must be used very frequently. For precise work, the heights of the various columns of mercury are read by means of a small telescope, known as a cathetometer, with cross hairs and a carefully graduated metal scale to measure the height. This method does away with the uncertainty as to the accuracy of the fixed feducial mark at the expense of somewhat more laborious calculations.

4. *Temperature Effects* We mentioned earlier the effects of temperature both on the dimensions of the various glass parts of the gauge and on the density of the mercury. These quantities can be taken into account by appropriate calculational techniques, but normally the room in which calibration is undertaken is maintained at a fixed temperature of approximately 23°C $\pm$ 3°C, which conditions are readily attained in practice and which do not induce appreciable errors due to temperature effects.

5. *Guidé-Ishii Effect* Guidé in Germany and Ishii in Japan have both investigated another form of error to which the McLeod gauge is prone and which has been named, after these two gentlemen, the Guidé-Ishii effect. This effect occurs because mercury from the gauge backstreams through the sidearm which connects the gauge with the system whose pressure is being measured. Under these conditions, the backstreaming of the mercury has the effect of causing the pressure in the gauge to be somewhat lower than the pressure being measured in the system to which it is attached and results in a fairly sizable error at the lowest pressures at which the McLeod gauge is capable of use.

6. *The Problem of Condensibility* The condensibility problem mentioned earlier exists in all McLeod gauges, and it is conventional to employ a cold trap in the line connecting the system with the gauge to remove water vapor and condensible materials from any gas passing from the system to the gauge. The same cold trap serves to collect and prevent mercury from the reservoirs of the McLeod gauge passing into the test dome to which the gauge being calibrated is attached, where it could seriously affect the reading of the ionization gauge by adding to the backstreaming effect discussed above.

7. *Long-pumpdown Effects* It should be emphasized that the McLeod gauge can only read the pressure which is present in terms of noncondensible gases in the test volume at the time the reading is begun and the volume cut off from the vacuum system. This pressure may or may not be the same as the pressure in the system. In general, it will not be the same as the pressure in the system if any outgassing effects are taking place in the McLeod gauge or if the time the McLeod gauge has been connected to the system at a particular pressure level has been insufficient for the pressure to equalize. The very nature of the McLeod gauge makes it necessary to use a rather long, high-impedance glass tube between the system being measured and the McLeod gauge itself. Periods as long as 24 hours are required to secure equilibrium of these pressures on initial pumpdown of the system. When readings are taken on a rising pressure curve, succeeding readings at higher pressures may be taken at much shorter intervals, but rarely at intervals of less than 4 hours if great accuracy is to be achieved. Readings taken at lesser intervals will always be suspect, since equalization of pressures through the long tubulation is slow and requires time.

11.6 *Use of the McLeod Gauge for Calibration Purposes*

Figure 11.2 illustrates one typical system for calibrating ion gauges against a McLeod gauge. The equipment shown has been taken from a specification on ion-gauge calibration of Committee E-21 of the American Society for Testing and Materials, which is charged with the task of developing pressure calibration methods for ion gauges. Basically, the test arrangement consists of a test dome of cylindrical cross-section,

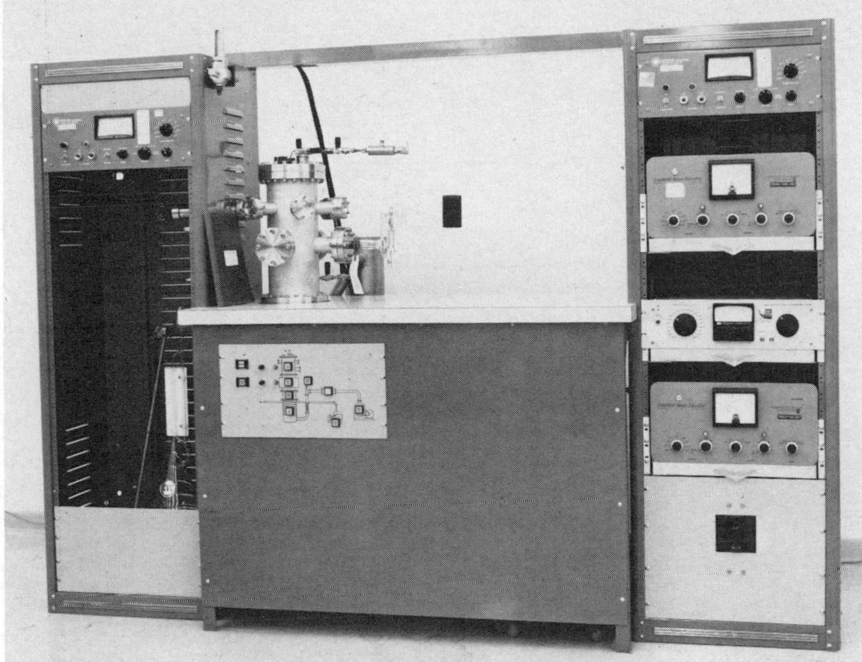

Fig. 11.2 Calibrating system for direct comparison with the McLeod gauge.

which may vary in general dimensions from that shown, mounted above a diffusion pump and well-cooled cold trap using liquid nitrogen for the coolant medium. All flanges above the diffusion pump should be of the metal-sealed bakeable type in order to permit baking the entire test dome above the cold trap. Inside the test dome there is provided a liquid-nitrogen-cooled finger which serves to condense any mercury backstreaming from the McLeod gauge before it can reach the gauges under test. The geometry according to which the McLeod gauge and the ion gauge face the cold finger must be identical, since temperature effects resulting from the reduced velocity of the molecules reflected from the cold finger into the gauge will affect the reading. It is important, therefore, that they affect the McLeod reading and the ion-gauge reading an equal amount, since otherwise the calibration would be invalid. The source of inlet gas to hold the system at the desired calibration pressure must be arranged in such a fashion as to prevent streaming of this gas into either the McLeod gauge or the gauges under calibration. In the example shown, the gas is brought in from the top in such a fashion that it directly impinges on the top of the cold trap and can only diffuse downward in a random fashion to the gauges. An alternate arrangement brings in the leak gas at the bottom just above the flange at the bottom of the test dome and directs it either upward again against the nitrogen finger to prevent streaming, or downward into the pump so that it is at an angle of 180° to the direction of the gauge connections. The gas is admitted through an adjustable calibrated leak which can be increased or decreased to provide pressure variation for running the calibration curve. The tubulation connecting the gauge to be calibrated and the test dome must be as short and large as possible; and, whatever the dimensions are, they must later be used in attaching the calibrated gauge to any test chamber on which it is to be used, as otherwise the gauge readings will be made erroneous by the effects of different tubulations.

11.7 *Accuracies of McLeod Gauge Calibration*

The accuracy of calibration made with the McLeod gauge as a standard depends largely on the absolute pressures which are being used. For pressures in the 1-micron (1×10^{-3}-torr) region, which might be used for calibrating thermal response gauges of the thermocouple type, accuracies of ± 10 percent of reading seem to be attainable. For the pressures in the 1×10^{-5}-torr region, where all of the effects mentioned above have full play, accuracies of better than 50 percent of reading seem doubtful. This type of work requires a precalibrated McLeod gauge, which is very difficult to achieve, as well as close control over all the variables mentioned above. In the 1×10^{-6}-torr range, the accuracy of the McLeod gauge method is such that errors of more than 50 percent of gauge reading can be expected, and the technique is therefore not of very much use for calibration work. If extreme care is used to control or compensate for all of the kinds of errors mentioned above, it would appear that accuracies in the 1×10^{-5}-torr range of ± 25 percent of reading can probably be achieved. However, to do so will require long and laborious procedures.

Where accuracies are mentioned above and in what follows, they are stated as percentage of reading within a decade. They further refer to the total errors of the calibration, including not only the inherent accuracy of the comparison gauge, but also reading errors, "streaming effect" errors, errors in the readout devices for the gauge being calibrated, temperature errors, and all other errors inherent in the calibration procedure.

Because of the various difficulties inherent in this method, and also because it is desirable to push the calibration limits downward into at least the high portion of the 1×10^{-8}-torr range, other methods of calibration are coming into use, and the McLeod method is gradually passing out of the picture. However, it is still useful for pressures in the 1-micron (1×10^{-3}-torr) or 1×10^{-4}-torr region, where conductance-limited types, next to be described, are not normally used.

11.8 *Conductance-limited Calibration Systems (Flow-based Systems)*

These systems, basically, make use of the relationship expressed by the simple equation $P = Q/S$, where P is the pressure, Q the quantity of gas being pumped, and S the speed of the pumping means. In such systems, large pumps connected to the system through relatively small orifices are used, and the systems are operated in the molecular-flow region only. Under these conditions, the pumping speed above the orifice can be accurately determined, either by calculation or by tests, so that one can determine definitely the speed of the pump and hold this constant over prolonged periods. The Q to which the test dome will be exposed will be that due to the inlet gas, which can be measured, plus that due to leakage, plus that due to outgassing of the chamber walls and tubulation. The leakage factor can be worked down to an extremely small value by meticulous care in manufacture and checkout of the system. The wall outgassing effects can be reduced by long, high-temperature baking and by stabilization periods of long duration at pressures 2 decades below the lowest test pressure to be used. Under these conditions, wall outgassing effects can be reduced to approximately 1 percent when pressure is raised the two decades to the test pressure. The problem then resolves itself into measuring accurately the quantity of inlet gas. Where quite low pressures are required, attenuation stages can be used, each connected to the other by orifices of known conductivity, and thus lower pressures may be obtained without restricting the amount of gas passing through the leak to unmeasurable values.

The simplest form of such a system is shown by the accompanying diagram, taken from the ASTM Committee E-21 Gauge Calibration Standard (Fig. 11.3). A photograph of such a system is shown in Fig. 11.4. Basically, the system consists of a stainless steel test dome in which any gauges to be calibrated, whether nude or tubulated type, can be inserted. Again, particular care must be taken to assure that the tubulation used during calibration is similar to that used on the system with which the gauge is eventually to be used, since otherwise tubulation errors will affect the readings. The system may be pumped by either a diffusion pump or an ion pump, whose total pumping speed is at least 10 times the speed of the leak orifice. This may be a calibrated conductance tube, as shown in the sketch, or a thin-plate orifice

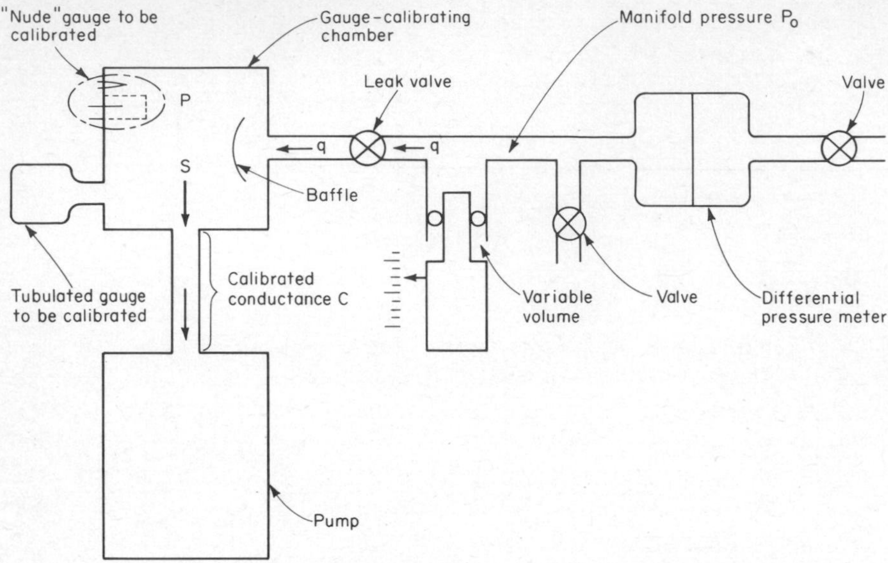

Fig. 11.3 Conductance for limited-calibration system (flow-based system).

Fig. 11.4 Photograph of conductance for limited-calibration system.

whose dimensions have been accurately measured by precise methods and whose conductance has been calculated using the Clausing correction values.

In order to assure that the speed of pumping in the test volume is strictly determined by the dimensions of the metering orifice, it is essential that the pressure on the pump side of the orifice be at least 1 decade below that of the test volume. To assure that it is indeed so, an ionization gauge must be provided below the orifice. Initially, the accuracy of this gauge cannot be known unless it has been calibrated on another system. However, after the first run, it can be replaced by the first gauge calibrated, if the accuracy of that gauge is known within at least 50 percent of reading. Successive recycling can then be used until the gauge below the orifice has been calibrated within acceptable limits.

Care should be taken never to operate the calibrating system at pressures where the 1-decade orifice differential cannot be reliably maintained. Serious errors will result from failure to observe this precaution.

In a typical system, as shown in the photograph, the pump speed is 1,000 liters per second and the orifice capacity is 100 liters per second. This means that a possible fluctuation in pump speed of 100 liters per second, or 10 percent, will be reduced to $\frac{1}{10}$ of this value, or 1 percent, in the actual test dome and will therefore introduce only a small error. The calibrating gas is admitted through an adjustable leak valve and directed against a baffle which prevents the streaming effects which would otherwise lead to nonuniform pressure response in the various gauges being calibrated. The gas is metered by means of a movable piston which is advanced in a cylinder filled with the calibrating gas at atmospheric pressure. A differential-pressure meter is employed to make sure that the pressure on both sides of the moving piston is the same so that leakage by the piston can be reduced to zero. The total amount of gas admitted is then determined by the actual volume of the metering cylinder and by the position of the moving piston. The position of the moving piston is read out by means of a dial-type micrometer of high accuracy. Normally, two or more pistons of varying diameter are provided to cover a large range of pressures more effectively.

11.9 *Multiple-conductance Methods (Pressure-based Systems)*

The principal difficulty with the system just described is that the measurement of the actual gas inlet quantities becomes very difficult at extremely low pressures. One way out of this difficulty is to use the

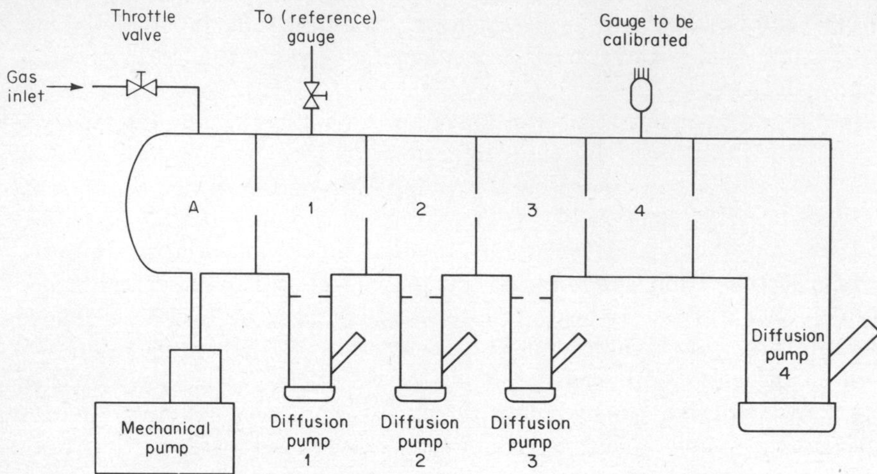

Fig. 11.5 Multiple conductance for limited system.

multiple-conduction system, basically shown in Fig. 11.5. In this system, it is not necessary to measure the quantity of inlet gas as long as the flow is adjustable through a very small leak valve. Instead, the pressure is attenuated through a series of chambers, each pumped by a diffusion pump with baffle whose speed is known and connected by a series of baffled openings whose speed is likewise known. With this type of system, the pressure will go downward in each chamber as compared to its predecessor, since the pressure on the orifice between the chamber and a succeeding chamber is continually going down. By proper selection, a gauge such as the McLeod can be attached to one of the early chambers where the pressure is sufficiently high to permit its use within its accurate range, and the pressure several stages downstream can be inferred from a knowledge of the various pumping speeds available and the number of stages.

11.10 *Calibration Methods in General*

Calibration of gauge tubes or sensors by the methods outlined above has led to a general tendency among those working in this field to go to conductance-limited systems for gauge calibration, since experience shows that at the point where the McLeod gauge and the conductance system intersect, the accuracy of the latter is somewhat better than the McLeod gauge accuracy, especially with respect to repeatability. It is also somewhat quicker and much less prone to trouble, since there is no need for baking out rather fragile glassware fo. lengthy times at relatively high temperatures. Impedance systems are all-metal and quite rugged in operation. They can also be baked out at higher temperatures and, in general, speeded up somewhat over the time necessary with the McLeod systems.

11.11 *Problems in Using Ionization Gauges*

An ionization gauge does not actually measure pressure, but rather measures the density or the number of ions that can be produced from the gas molecules within its grid cage. If the calibration is correct, it will therefore read with a reasonable accuracy the pressure within the gauge itself. However, this pressure may or may not be the same as the pressure within the system being measured. Tubulation effects of the connection between the gauge and the system are quite great and, in many cases, cause the gauge to operate either at a higher or a lower pressure than the system, depending on whether the gauge is basically outgassing or acting as a pump, both of which it can do under certain circumstances.

In addition, the tubulation of the gauge acts as a filter permitting only those ions to enter the gauge, and thus be counted and measured, which approach the gauge tubulation at an angle roughly parallel to the axis of the tubulation. Molecules that happen to approach the mouth of the tubulation on the vessel side at an angle too great will, of course, impact the walls of the tube but will then be reflected outward again and will not get into the gauge-tube cage to be read. The gauge therefore has a strong directional effect, measuring basically that portion of the ion population in the chamber which happens to be moving in line or nearly in line with the gauge tubulation.

In actual systems, the motion of molecules within the chamber is not fully random, due to the fact that pumps are attached at some point and are removing molecules in this region while gas coming from work being done within the chamber is being emitted in nonrandom distribution from the work area. There therefore exist definite streams of molecules within the chamber which are quite nonrandom in orientation. The positioning of a tubulated gauge, therefore, determines to a large extent the pressures which the gauge will see and thus read out. Differences within large space-simulation chambers containing outgassing loads which are irradiated by simulated solar energy and surrounded by cryogenic surfaces have yielded differences of pressure within the same chamber at the same time as high as a decade simply due to this nonuniform streaming effect due to outgassing within the system combined with pumping capacity in specified locations.

In an exploration of this kind, tubulated gauges are frequently placed inside the chamber or mounted on the chamber walls in specified locations such that the streaming effect can be utilized to determine the place from which gas is coming and the relative efficiency of various pumping means in removing it. It may also be that the point of interest is the effective percentage of those molecules leaving the test specimen which return thereto after being reflected from the wall. In such cases, the gauge may be so placed as to face away from the source of the gas and toward the pumping means or the walls of the chamber, in order to measure not the flux of gas from the specimen but rather that portion of this flux which returns to the specimen after being reflected.

Where the problem is simply to measure the average chamber pressure as accurately as possible, tubulation that is as short as possible and as large in diameter as practicable must be used so that the effect of the tubulation can be reduced to a minimum.

A more effective way of determining average pressures within the chamber is to use a nude gauge without encapsulation directly exposed within the chamber.

Unbaked tubulated gauges take very long periods to stabilize. In some work performed by the author and his associates, it was found that: (1) At a pressure of 5×10^{-8} torr, it required 30 hours with the filament on for an unbaked ion gauge with 0.43-inch inside diameter (I.D.) tubulation to reach equilibrium (no change in reading as a function of time and at a constant pressure); (2) at pressures of 1×10^{-7} torr, it required 6 hours for the ion gauge with a standard tubulation (0.55 inch I.D. for a triode tube and 0.43 inch I.D. for the Bayard-Alpert type) to reach equilibrium; (3) a gauge with 1-inch I.D. required approximately 2 hours to reach equilibrium at this pressure. For measuring pressures below 1×10^{-7} torr, either a very large tubulation on an unbaked gauge or a gauge with normal tubulation plus baking is required. Even with the nude-type gauges, some form of degassing by bombardment or baking must be employed to secure correct readings.

It must be remembered that the conductivity of a given tubulation varies inversely with the square root of the molecular weight of the gas molecules being handled. The use of tubulation, therefore, has a separation effect which tends to make the composition of the mixture of the gases within the gauge different from that within the chamber.

All of the hot-filament-type gauges have x-ray limit effects which limit the minimum pressure which they can read. In the best of the Bayard-Alpert types, the gauge reading generally becomes questionable somewhere in the neighborhood of 2 to 5×10^{-10} torr. If the pressure in a system is determined by some other means, such as a cold-cathode gauge which is relatively unaffected by the x-ray phenomenon, it will be found that, in general, the Bayard-Alpert type gauges will begin lagging behind the cold-cathode gauges at pressures on the order of 5×10^{-10} torr, but that if sufficient time elapses, frequently 24 hours or more, the Bayard-Alpert gauge will in time read down into the 1×10^{-11}-torr range. This would seem to indicate that the phenomenon causing the so-called x-ray effect is only partially due to x-rays, which would of course be constant with time, and is at least partially due to some reactions at the hot filament or other gauge parts which limit the ability of the gauge to reach lower values. With time, these reactions cease and the gauge reads lower, being eventually limited by the true x-ray current effect.

11.12 *Outgassing of Gauges*

Most commercial Bayard-Alpert gauges have some provision for internal outgassing. This may be done by a special circuit in the power supply which causes a heavy current to pass through the grid and collector, in order to heat these parts to a high temperature to outgas them. More frequently, outgassing is done by superheating the filament and applying a high positive voltage to both the grid and collector, thus bombarding them with electrons to produce heating. Alternately, the entire gauge, including the envelope, may be heated by an external oven or by external heaters.

Any of these methods will cause outgassing of the gauge if carried on for a sufficient time. However, the freshly outgassed gauge will act as a pump for some considerable period of time. It will therefore give rise to erroneously low readings. This may make the operator happy, but contributes nothing to the validity of the work. The "Ion Gauge Application Standard" of ASTM Committee E-21 calls for not less than 2 hours between the outgassing operation on a tubulated ion gauge and its use as a pressure measuring device. Frequent "degassing" of the gauge—before each reading, as is sometimes done—can only give a series of erroneous readings on the low side. The writer's own preference, when forced to use a tubulated gauge, is to do a degas operation by electron bombardment as soon as the system has reached about 1×10^{-5} torr, and thereafter do no further degassing during the run.

Of course, all these problems can be substantially eliminated by the use of nude ion gauges within the chamber. The nude gauges also eliminate another annoying effect (the Blear effect) caused by the condensation of a monolayer of oil-vapor hydrocarbon molecules on the inside of the tube envelope. In addition to the conductance effects due to the flow of this vapor into the tube envelope, the oil layer is extremely temperature-sensitive. Very slight changes in temperature, due to the touch of a hand, for instance, can cause large effects in the indicated pressure.

Nude gauges not only have the advantages outlined above but are nearly undirectional, and thus give a much less biased indication of chamber pressure than any tubulated gauge, unless static conditions exist throughout. Generally, at pressures of 1×10^{-5} to 1×10^{-6} torr, the nude gauge will read higher than a tubulated gauge. At 5×10^{-10} torr, the nude gauge will read lower than the tubulated gauge.

chapter 12

The Vacuum Vessel

12.1 *Introduction*

Most of the troubles which occur in vacuum systems as set up in test laboratories, in environmental test chambers, or in vacuum coaters occur due to leakage in the vacuum vessels or in the penetrations of the shell for various gauges and passthroughs. This is true because the methods of fabrication ordinarily used in fabricating mechanical joints or joints for pressure vessels are not adequate to produce vacuum-type joints suitable for high or ultrahigh-vacuum systems. Special techniques are therefore necessary in making such joints which vary with the materials being joined. Vacuum vessels of various types can be and are being made of various materials, including glass, copper, brass, carbon steel, aluminum, and stainless steel. Each material requires a somewhat different technique for its fabrication, and the details of these techniques will be discussed in the following sections.

12.2 *Mechanical Considerations*

There is no standardized code covering the design of vacuum vessels. The American Society of Mechanical Engineers' "Unfired Pressure Vessel

Code," which is very widely used in industry, specifically exempts vacuum vessels from the requirements of this code. Nevertheless, mechanical considerations must be given due weight in designing the vessel. It is true that the maximum pressure to which a vacuum vessel can be subjected is 15 pounds per square inch, the weight of the atmosphere. However, this force is directed inwardly at all points instead of outwardly as is customary in pressure-vessel work. The chief danger is a collapsing action, in which one section yields to the forces involved and buckles inwardly, leading to widespread collapse of the whole vessel.

Most vacuum vessels are circular in cross-section in one dimension and have domed heads. Rather simple formulas are available from textbooks and handbooks on mechanical design by which the stresses inherent in a structure of this nature can be readily calculated. Care must be exercised, however, to apply these formulas with a considerable amount of caution, since a departure of the vessel from true circular cross-section by only a small amount greatly increases the collapsing forces; and rolled shapes fabricated by the usual techniques are seldom precisely circular. For this reason, eccentric loads of reasonable amounts should be allowed for, and the thicknesses of the plate and reinforcing structural material should be calculated to allow for these forces and to restrain them adequately. Several cases have occurred where vessels, otherwise adequate, were insufficiently reinforced in the region of a large penetration or door which entered through the cylindrical section, resulting in collapsing forces sufficient to distort the whole region of the vessel surrounding this penetration with consequent large costs involved and serious delays in the projects which were to use the vessel. Particular attention should be given to a stress analysis around large penetrations and adequate reinforcing material provided to take up the loads at such points.

12.3 *Glass*

Glass is quite frequently used for the construction of small laboratory vacuum systems used for test purposes and, if properly handled, can produce very satisfactory results. Glass chosen for the assembly of such systems should be high-quality Pyrex glassware having a low thermal coefficient of expansion. If possible, large sections should be annealed following fabrication or, in the case of bell jars, annealing carried out before the system is assembled. Joints, wherever possible, should be formed by means of glass welding using a suitable oxyhydrogen torch. The technique employed must be as perfect as possible, since joints that appear to be perfect can easily include small open stringers which lead to relatively large amounts of gas inleakage, which make the attainment of desired vacuums difficult or impossible. Such types of leaks are very difficult to detect by ordinary means and are almost invisible to the eye. Frequently, the only method of readily detecting such situations is by use of the Tesla coil, where a high voltage is applied between the inside and the outside of the glass system. The presence of such a small leak will become instantly visible due to the conduction of a high-voltage streamer through this point. When the leak is found, repair can sometimes be made by the use of a torch, but often the only way to eliminate the difficulty is to replace the defective piece of glassware with a fresh piece. Unfortunately, such defects are sometimes present in as-received glass stock or glass parts and are invisible until the system is welded into the structure and a vacuum pulled, after which the usual checking techniques or the Tesla coil can sometimes locate them.

Where movable joints or joints which can be taken apart must be made, ground-glass joints coated with a *very thin* layer of vacuum grease can be used in relatively high-pressure vacuum systems. For extremely low-pressure systems, a joint using one of the meltable sealing waxes must be used in order to assure tightness. Unfortunately, most of these waxes have the property of hardening after prolonged use, resulting in the appearance of very small cracks which admit air to the system and which, again, are difficult to locate and correct. So far as possible, therefore, joints that are not welded should be avoided in the system, even if cleaning necessitates cutting such a welded joint and rewelding it after the cleaning is completed.

It should be borne in mind that Pyrex glass, which is preferred for glass systems because of its low coefficient of expansion and strength, is permeable to some extent to helium, which is a constituent of the atmosphere; and we can therefore expect that, in a glass system, slow inleakage of helium will be encountered in the ultrahigh-vacuum regime, which must of course be removed by the pumps. For this reason, ultrahigh-vacuum systems are usually fabricated of stainless steel, which is much less permeable to helium than glass.

12.4 *Copper and Brass*

Both copper and brass can be used in vacuum systems, but brass is to be avoided in any system which must reach high- or ultrahigh-vacuum levels, particularly if any heat is to be encountered in service or during bakeout. The reason of course is, as mentioned in Chapter 3, that the zinc content of the brass slowly vaporizes from the vacuum side of the vessel, resulting in contamination of the system.

Copper is quite satisfactory for vacuum-vessel assembly and is widely used for small systems. It can be fabricated by soldering or brazing joints in the parts being assembled into the system. In general, soft tin-lead solder should be avoided for vacuum systems since it has certain serious weaknesses. At low temperatures, which may be attained near liquid-nitrogen cold traps, the soft solders tend to become brittle and crack due to a phase change in the tin present, causing leakage. At temperatures attained during bakeout, which is necessary to clean up the occluded gases in ultrahigh-vacuum systems, the solder would, of course, melt and therefore cannot be used. At temperatures below the melting point but above room temperature, tin tends to vaporize out of the solder, depositing on the interior of the system and sometimes shorting electrical contacts. This is not a serious problem if temperatures remain at or near room temperature.

The most satisfactory medium for joining copper parts in a vacuum system is silver solder-type brazing material, melting at approximately 1400°F. Unfortunately, the use of silver solder-type material necessitates the use of a flux to enable the soldering material to make a tight joint with the mating surfaces. If any of this flux remains on the vacuum side of the system, it will contribute large amounts of gas to the vacuum chamber, resulting in very poor performance of the system, which can continue for a very long period, since the residual flux will pick up additional water vapor every time the system is opened.

It is therefore desirable when using silver solders for joining copper systems to make sure that the technique is such that the silver solder thoroughly fills the joint and that any flux which could contribute porosity or other difficulties to the joint is washed out of it. It is then necessary that all traces of the flux be thoroughly removed from the interior of the vacuum system by hot-water washing and otherwise cleaning the system after the joint has been completed. This operation is difficult, since frequently the solvents of the flux will not reach certain corners or cracks in the joint and therefore leave behind some residual flux. Nevertheless, silver solder is widely used for many vacuum joints and is satisfactory if good techniques are used in applying and cleaning the joints. Cadmium-bearing silver solders should be rigorously avoided, due to the high vapor pressure of cadmium.

12.5 *Carbon Steel*

High-quality carbon steel, such as boiler plate, can be used for the fabrication of high-vacuum systems. In using such materials, it is extremely important that all oxide and scale be removed from the interior portions of the vessel after joining, and that these also be removed from areas to be welded before welding is carried out. The surfaces exposed to high vacuum must then be thoroughly cleaned of all oxide and contaminants of whatsoever kind and normally given a cleanup treatment by sandblasting, grinding, and/or polishing. With this preconditioning, a reasonable pumpdown performance can be expected. However, the principal difficulty with carbon steel vessels is that each time the clean inner surface of the vessel is exposed to room air, carrying considerable quantities of water vapor, a small amount of rust will form on the inner surface. This rust coat, which may be so thin as to be invisible to the eye, will then pick up and hold great quantities of water vapor which are removed only with great difficulty or long pumpdown time on the next round of pumping. As a result, the use of carbon steel in vacuum vessels, where reasonably low pressures are required, is decreasing and this material is now used only for large vessels that are to be used only for a very limited campaign and then scrapped. The savings in pumpdown time through the use of stainless steel amply repay the additional cost if long service is to be expected.

In small vessels it is sometimes possible to avoid the rusting effects mentioned above by applying an evaporated coating of aluminum followed by an evaporated coating of silicon monoxide (which converts to silicon dioxide upon exposure to air). If relatively thick coatings are applied (to eliminate pinholes), performance closely approaching that of a stainless steel vessel can be obtained. The principal drawback is that if a later cleaning operation is required (as it surely will be, sooner or later), the coating will quite possibly be destroyed by cleaning, thus requiring renewal to restore the protecting film.

12.6 *Aluminum*

Aluminum can be and is used for high-vacuum systems and, if properly cleaned prior to use, is an entirely satisfactory medium. The joining of aluminum parts in a vacuum-type weld is more difficult than joining equivalent parts of steel or stainless steel, but can be achieved with proper care in the welding technique, using tungsten–inert-gas (TIG) welding techniques (see Chapter 13 for a discussion of TIG welding).

When aluminum is used for a vacuum vessel, care must be taken to choose one of the aluminum alloys which exhibits good welding properties, as otherwise porous welds will result which will not show the zero leakage required for vacuum welding. In general, the high-strength aluminum material, such as 2024, 7075, and similar heat-treatable alloys, are not suitable for vacuum vessels, since the alloy elements tend to give rise to porous welds which are not permissible. However, the aluminum alloys specifically designed for welding, such as 6061, 3003, and 5054, are suitable for vacuum-vessel construction. For cryogenic vessels, 2219 is preferable although somewhat more difficult to weld than the other materials.

12.7 *Stainless Steel*

Materials listed heretofore are suitable for vacuum vessels that need not be baked to a high temperature for outgassing purposes. For ultrahigh-vacuum operations, however, where baking is mandatory in order to clean up the surfaces of the vessels sufficiently well to permit the achievement of these ultralow pressures, the use of aluminum, copper, etc., is substantially prohibited by the bakeout requirements. If any baking is to be done on vessels of these materials, it must be at an extremely low temperature, which is not particularly effective in eliminating the outgassing effects to the degree necessary for ultrahigh-vacuum work.

For this reason, among others, stainless steel has become the principal material for the construction of all vacuum vessels having requirements of 1×10^{-6} torr and below. The preferred materials are AISI 304 and 321, although several other of the stainless alloys are acceptable if desired. These materials can be welded by several processes, of which the tungsten–inert-gas (TIG) method is the most useful for vacuum work. They are strong and readily resist the pressures inherent in vacuum-vessel design, even at the elevated bakeout temperatures. They also reduce oxidation and thus will stay clean internally much better than the other materials. They readily accept sandblasting, grinding, and polishing techniques to produce a smooth internal surface in order to reduce the amount of gases occluded thereon and contributing to outgassing effects. Except in very special and unusual cases, all parts of the system, including the heavy door flanges, penetration flanges, etc., should be fabricated of stainless steel of approximately the same alloy in order that the coefficients of expansion may match and to facilitate closures that must be applied to these penetrations.

Cast materials are not generally satisfactory as primary vacuum structures. The difficulties arise from two possible types of deficiencies. In the first place, castings often exhibit slight porosity, which gives rise to leaks that are very difficult to locate and almost impossible to correct. This has occurred in several aluminum permanent-mold castings for vacuum gate valves, which leaked unacceptably when bagged but exhibited no locatable leaks. Such castings must either be discarded, or some sort of impregnating compound used which will close the porosities. This latter method is only effective if the leak is very small, and if no heating is to be employed which might act to destroy the impregnating material.

The second difficulty arises from the relatively rough as-cast surface, which acts to hold on to very large amounts of moisture, grease, and contaminants, rendering outgassing very high. This effect can be avoided by machining all vacuum-exposed surfaces—again provided no microporosity is present to defeat the effort.

chapter 13

Welding for High Vacuum

13.1 *General Requirements*

The importance of welding techniques in the assembly of high-vacuum systems is so great that it has been felt desirable to discuss in some detail the welding methods which have proved successful. Variations in this technique can, of course, be made at the discretion of the user; but changes to various "cheaper" welding techniques such as coated-electrode arc welding have generally resulted in not only having to grind out and repeat the weld using the proper techniques, but great expenditure of time and money to accomplish a job that would have been simple had proper techniques been used in the first place.

In a vacuum system, two things are paramount for the production of successful welded joints: (1) an experienced welder, and (2) correct design. The theme of the design should be joints so designed as to permit 100 percent penetration, with no trapped volume on the high-vacuum side.

Inclusions and segregations in the steel in the vicinity of the weld can produce microcracks and porosity open to the vacuum side. Even though these may not appear to pass completely through the section,

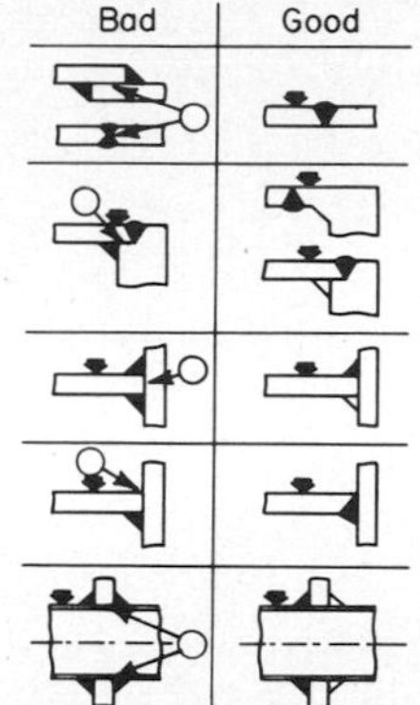

Fig. 13.1 Comparative utility of various vacuum-welding practices.

it has been found from experience that gas penetration almost always shows up sooner or later in the vicinity of any such inclusions or microporosity areas. Even when working at a pressure of 1 micron (1×10^{-3} torr) there is an expansion ratio involved of 760,000 to 1. Very small amounts of gas, therefore, coming from trapped volumes or small porous spots, can prevent a system from reaching the desired level of vacuum. Experience has indicated that the only completely satisfactory welding method suitable for high-vacuum work is the inert-gas-shielded tungsten-arc process. This process is called tungsten–inert-gas (TIG) welding by the American Welding Society. Other methods utilizing coated rods, submerged arcs, etc., have not been able to produce leak-free joints in high-vacuum enclosures. Worse yet, joints once so welded cannot be satisfactorily repaired unless the entire weld is removed and rework is by the inert-gas-shielded tungsten-arc process. It is therefore imperative in attempting to do fabrication of hard-vacuum systems that the inert-gas-shielded tungsten-arc (TIG) or metal–inert-gas (MIG) welding process using a consumable *bare* electrode be used. Any attempt to use any of the faster and cheaper welding processes will only yield in the end more work and more cost than if the proper method had been used in the first place.

Figure 13.1 illustrates proper and improper methods of welding. In general, welding for high-vacuum systems should be done, so far as possible, from the vacuum side. By this method, the underbead porosity and inclusions inevitable at this point are not exposed to the hard vacuum. In all cases, inert-gas protection must be provided both on the underbead and on the welding side of the joint in order to prevent the formation of tenacious oxide films which give rise to leakage or porosity.

The designer must detail the edge preparation to assure 100 percent weld penetration, suitable designs being given in Figs. 13.2 and 13.3.

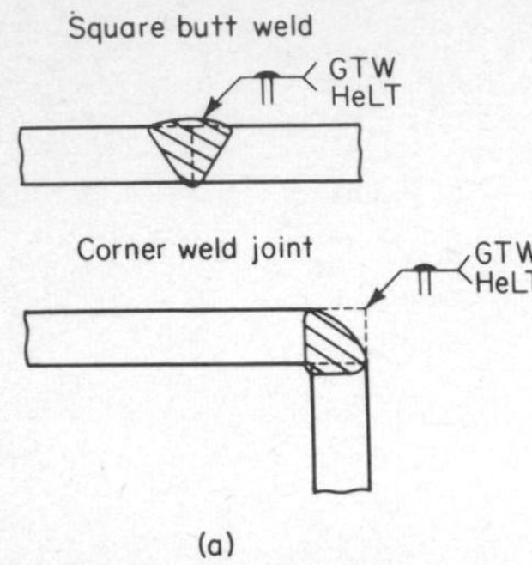

Fig. 13.2a Weld preparation, materials up to 0.100 inch thick: edge preparation for butt and corner weld.

The welding symbol tells the welder which side to weld from and what process to use. The note in the tail of the symbol can include the filler metal when required and the statement of integrity. Two typical symbols are shown in Fig. 13.2.

After welding is complete, each section of the system should be leak tested by the helium mass-spectrometer method wherever possible before additional sections are added. Where leaks occur, the leaking section must be ground out completely and the surfaces recleaned, rewelded, and retested. It is emphasized again that coated-rod welding beads are not repairable except by completely machining away the total weld bead and remaking the weld by the tungsten-inert-gas (TIG) process.

13.2 *Welding Rods*

For 321 or 347 stainless steels, AISI 347 welding rod has been found most satisfactory, although 308 ELC rod is a good alternate, especially for joining 304 to 304 or to 321. Where welds must be made in mild structural steels, the only satisfactory welding rod found so far is Oxweld 65, a triple-deoxidized material which seems to give very low-porosity welds. Other rods than these may be useful, but experience indicates

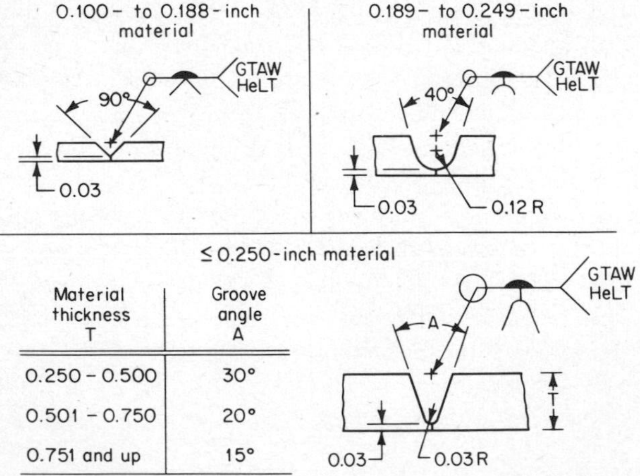

Material thickness T	Groove angle A
0.250 - 0.500	30°
0.501 - 0.750	20°
0.751 and up	15°

Fig. 13.2b Weld preparation, materials over 0.100 inch thick: butt-joint designs and drafting symbols.

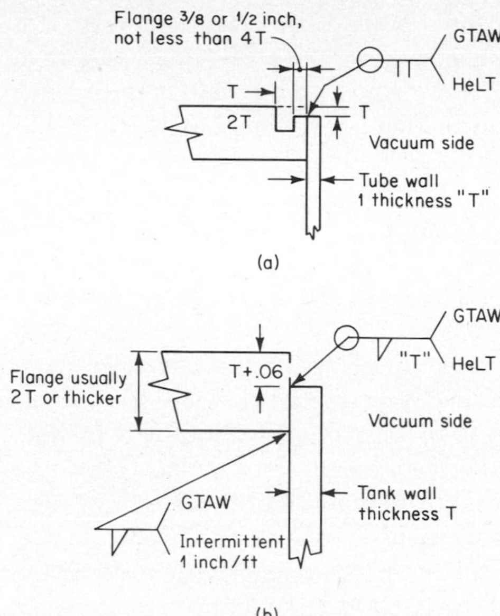

Fig. 13.3 Joint design: edge preparations for welding flanges to (*a*) tubing and (*b*) tank wall.

that porosity in the weld invisible to all inspection methods except helium leak tests is more apt to occur; and they should therefore not be used in high-vacuum systems.

The filler rod must be purchased clean and kept that way. For highest-quality work, it will be wiped clean and dry just before use with acetone or methyl-ethyl-ketone (MEK) and oil-free gauze.

13.3 *Welder Qualifications*

Welders who are to work on systems to be used under high- or ultrahigh-vacuum conditions should be qualified before starting work on the project, even though they are thoroughly experienced in other aspects of welding. A good technique to qualify the welder is to have him weld a test assembly consisting of two pieces of rolled-up stainless steel which must be seam-welded, joined at 90 degrees, and provided with a flange at one end to connect to a pump system, pressure gauge, and leak detector and a blind closeout weld at the other end. Figure 13.4 outlines one typical configuration that has been successfully applied to this purpose.

When the welder has finished the test piece, it is then connected to a vacuum system, a pressure gauge, and a leak detector of the helium mass-spectrometer type and pumped down. When leaks are discovered (as they invariably will be), the welder is called upon to grind out the welds and repair them until a tight system has been achieved. Initial tests can be made by simply blanking off the system and observing the pressure rise, if any. When the system no longer rises rapidly after blankoff, the helium leak detector should be applied. This technique is guaranteed to bring out to the welder the differences between vacuum welding and other types.

The time involved in making up the test piece and repairing leaks that occur in it will be more than saved in actual welding of the vessel to be fabricated. Generally, it takes about two days for an otherwise skilled welder to learn the peculiar techniques required for high-vacuum-system welding. However, if qualification is not carried out first, a much longer time will be spent in repairing the actual vessel when the inevitable leaks show up in the larger fabrications. It is therefore time well invested to train skilled welders in the peculiar techniques of tungsten–inert-gas (TIG) arc welding as required for vacuum systems.

Of course, the small qualification assembly may be used to test edge preparation, material thickness, and alloy combinations to be used in the finished vessel as well as to develop welding and repair techniques.

It is worth emphasizing that the cleanest materials possible should be procured for hard-vacuum vessels and strict cleanliness maintained in and near all abutting surfaces if rework of the welding is to be held at a minimum.

The design of the test vessel shown in Fig. 13.4 includes butt, corner, and fillet welds encountered in most actual hard-vacuum-system components.

13.4 *Temptation*

Since the TIG welding process is commonly considered the most expensive welding process, the temptation is strong to substitute other processes or combinations. One such often suggested for use on ⅜-inch and thicker materials is to TIG-weld the vacuum seal pass, then fill the joint by the submerged-melt or the manual coated-rod process. Experience has proved that this does not work. The welding fluxes penetrate the TIG-deposited metal in a form of intergranular corrosion, producing copious leakage. Attempts to repair the outer passes by TIG or any other welding process may eliminate the leaks that were ground completely out, at the same time generously opening up new ones adjacent to the repair, and so on ad infinitum.

A more successful technique, which is workable for plate ½ inch or more in thickness, is to weld through the joint by coated-electrode welding, check the joint for porosity by x-ray techniques and careful visual inspection, then grind off the inner surface of the weld to reach

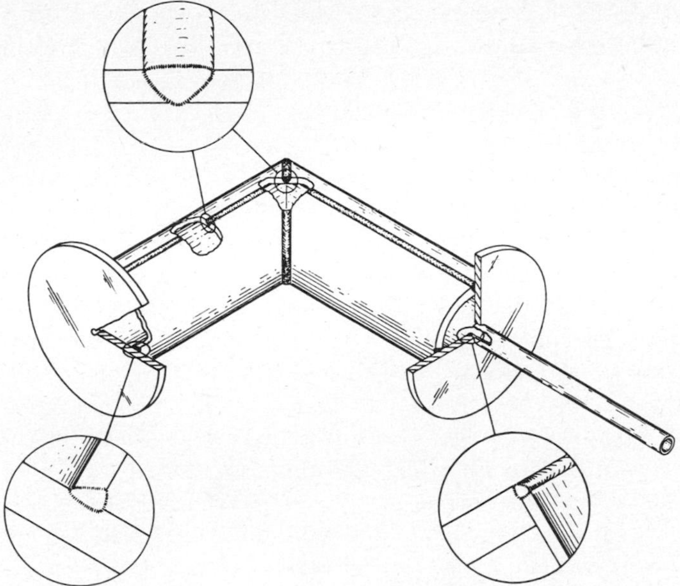

Fig. 13.4 High-vacuum-qualification welding test assembly.

dense, solid metal, and do a good wash coat with the TIG process on the high-vacuum side of the weld. With this technique, such flux inclusions as exist will be in the outer portions of the weld and not exposed to the high vacuum, with the inner surface being sound due to grinding of the first weld layer to sound metal and then wash-coating with a good TIG coating.

13.5 *Weld Preparation and Technique*

Figure 13.1 shows various welding practices useful in vacuum welding, together with similar practices differing in some respects which are unsatisfactory for vacuum welding. The defects resulting from these unsatisfactory techniques involve the creation of unsealed volumes containing gas, slag, and dirt which can appear wherever a minute crack exists in the inner weld, even some distance from the pocket in which most of the dirt is accumulated. Such virtual leaks are very difficult indeed to locate and almost impossible to eliminate if they are allowed to occur. In general, continuous welding on both sides of any joint is to be avoided like the plague in vacuum work. A continuous weld on the high-vacuum side is preferable for all possible applications. If additional weldment must be deposited on the outside of the joint for reasons of strength, this should be intermittent rather than continuous, thus permitting the application of the helium tracer gas to the seal weld and the location of leaks, if any exist, in it. If this is not done, it becomes impossible to locate the minute cracks or leaks in the inner weld with any precision.

The joining of a heavy flange to a thinner tube or sidewall section may be accomplished by either of the techniques shown in Fig. 13.3. Figure 13.3*a* shows a configuration conventionally employed for bakeable flange connections where a meltdown without the addition of filler material is employed to make the seal. The relatively thin sections allow a reasonable control of heat during the welding operation to prevent warpage or undue locked-in stresses and to give some degree of flexibility to the resulting structure.

The arrangement shown in Fig. 13.3*b* is more usual where heavy door flanges must be joined to sidewalls. In this case, added weld-filler material is, of course, necessary. The symbol "GTAW" used to describe the weld stands for gas-tight argon-shielded tungsten weld. The second symbol, "HeLT," calls for helium leak testing of the weld.

The following notes apply to all vacuum welding where high performance is required.

1. All welding in hard-vacuum systems requires 100 percent penetration and 100 percent inspection by the helium mass spectrometer.
2. The drawing shall include a section view detailing the edge preparation in thickness above 0.100 inch.
3. Separate backup strips and consumable inserts are not recommended in hard-vacuum systems.
4. Welding from one side of the joint gives more satisfactory results and easier isolation and repair of leaks than welding from both sides.

13.6 *Welding for Ultrahigh-vacuum Systems*

The welding requirements for systems that must achieve pressures below 1×10^{-9} torr are basically similar to those described above. In this case, however, there is no alternate for the use of stainless steel, and the welding techniques employed must be as near perfect as is humanly possible. Not only must the resulting vessel be extremely tight, dictating the ultimate in weld preparation and in care in welding, but it is also highly important that the welds be free of virtual leakage through porosity, dirt inclusions, and the like, which is an ultrahigh-vacuum system can be catastrophic. When more than one welding pass is required, it is usually necessary, even with TIG welding, to grind the surface of the early passes before additional passes are applied on the vacuum side, thus removing any oxide that may have formed near the top of the first melt, even when protected with shield gas as is mandatory for these systems. Thus the final melt will be made on top of solid metal with no dirt or inclusions visible to the naked eye, even with the most careful inspection. Shielding gas must be applied on both sides of the weld in all areas where hot metal exists. Argon has proved quite satisfactory as a shielding gas in place of the more expensive helium.

After the welding is completed, it is necessary (for ultrahigh-vacuum systems) to grind the vacuum surface of all welds, inspecting them carefully for porosity, dirt, etc., and to grind and/or polish all internal surfaces exposed to the high vacuum. Surfaces that are cleaned by pickling or sandblasting can be used and will, in general, permit the achievement of pressures in the 1×10^{-10}-torr range or lower. However, the relatively rough surfaces are easily contaminated during later use when specimens or operations performed within the chamber can be expected to liberate relatively large quantities of contaminating material. These rougher surfaces are much more difficult to clean for the next run than the polished surfaces which are recommended. It has been found, therefore, that the cost of doing a full polishing job on the interior surface is well worthwhile, since the cleaning after subsequent contamination is so much easier that the cost of the polishing operation is repaid the first time recleaning is required. In addition, since the surface area of a polished piece of metal is somewhat less than the surface area of a sandblasted or pickled piece, the amount of outgassing even on the initial run will be somewhat less.

Wherever possible, plate should be ground and polished before forming and fabrication. For large vessels, stainless steel plate can be procured with one surface ground and polished to "simulated No. 6 mill finish." The surface is protected by paper during shipment and fabrication. This technique greatly reduces the cost and labor required for final cleanup after fabrication.

chapter 14

Closures

14.1 *Introduction*

In what has been said heretofore, we have discussed the fabrication of tight vacuum systems without openings. However, in any system which is actually to be used, one or more openings are essential, arranged so that they can be opened or closed relatively easily, either to allow insertion of materials for operations in the vacuum or to permit adjustment of internal parts. We shall discuss several means of making such closures and the pitfalls that exist in the assembly techniques. It will be necessary to discuss these matters under two headings—one dealing with medium- and high-vacuum systems suitable for use into the high 1×10^{-8}-torr range, the other with ultrahigh-vacuum systems useful at 1×10^{-9} torr and lower—since the techniques differ in the two pressure regimes.

14.2 *Pipe Threads*

Pipe threads are nearly always necessary for thermocouple gauges and may be required for ion gauges and other types of penetrations, especially in connection with the roughing systems of vacuum chambers.

The difficulty of making tight threaded joints is directly proportional to the diameter of the pipe being joined. In general, the small pipes used for gauge tubes give little trouble in the medium and high-vacuum range if properly made, but threaded joints larger than 1 inch are nearly impossible to maintain tight.

Any pipe threads to be used in a vacuum system must be very carefully cut with new, unworn dies. Threads should be sharp and clean, without tearing or cracking of the thread area, which is common with worn dies. If this occurs, it will be impossible to secure a tight closure of the threaded joint.

Pipe threads, and indeed all types of thread systems within the vacuum system, can be the source of very serious virtual leaks due to air trapped in the thread area. The best technique for closing such joints is to make use of Teflon tape, which is a thin tape made of pure Teflon somewhat similar to Scotch tape in appearance. A single layer of such tape is stretched tight around the threads and then forced into the mating thread at high tension. Under these conditions, the Teflon distorts to fill the thread area and provide a tight seal. If the joint must afterwards be removed, removal is relatively easy and the tube can be replaced by using a new section of Teflon tape after carefully cleaning out the remains of the previous one, which will have been destroyed.

It is important to recognize that no type of pipe dope or compound for threads known to the writer can be used successfully on threaded vacuum joints. Glyptal lacquer has been used in the past, but is subject to aging and cracking, giving rise to leaks where none originally existed and making said leaks very difficult to find. For this reason, the use of the Teflon tape is becoming universal in vacuum practice.

It should be obvious that, before making up the threaded joint, the freshly cut threads should be thoroughly washed and cleaned in acetone or methyl-ethyl-ketone (MEK) and carefully dried before applying the tape and making up the joint, since trapped oil will give rise to endless difficulties.

Threaded joints, because of their tendency to develop leaks, are normally used only in forepump lines where the pressure is relatively high or in gauge connections where the size is small.

In ultrahigh-vacuum systems, threaded joints of any variety are completely inadmissible, and passthroughs for this type of system must be made by flanging with bakeable metal-sealed flanges.

14.3 *O-ring Seals*

The most common type of seal used for closing vacuum systems is the O ring. However, the use of O rings for closing high-vacuum systems differs somewhat from the methods of applying such rings for pressure systems.

In order for any O ring or other gasket material to be suitable for high-vacuum service, it must have a very low vapor pressure at the service temperature; that is, it must not have any components which will vaporize at the pressures and temperatures to which the joint will be subjected. In addition, it must have a very low permeability; that is, gas must not readily pass through the bulk body of the sealing material. Unfortunately, these two properties do not go hand in hand, so that frequently we have materials which have very low permeability but which have a high outgassing rate, whereas other materials with low outgassing rate have relatively high permeabilities.

In all cases, the vapor pressure of most organic materials is inversely proportional to the temperature. It is therefore very important that O rings be maintained at all times as cool as practicable, even when the system is being baked out. The best of the materials must be held below 250°F, and for minimum outgassing an even lower temperature is desirable. In fact, if the temperature can be held at 45 or 50°F, it is possible to use carefully constructed O-ring joints at pressures down as low as 1×10^{-9} torr.

In general, neoprene is the most commonly used material for gaskets, being used at a Durometer A-scale hardness of 60 to 70. Teflon is somewhat better than neoprene but has creep or cold-flow characteristics which may require periodic retightening of joints in which it is used. Some of the silicone rubbers have been used, but these normally have not been as low in permeability as some of the other materials.

The best of all the O-ring materials currently available is Viton-A (DuPont), which can be used down to 1×10^{-9} torr. It will also withstand mild bakeout, provided it does not reach temperatures higher than 250°F.

Probably the lowest-permeability material available is butyl rubber, which of course has been used for this property in automobile tires for many years. Unfortunately, this material has a high outgassing rate and therefore, if used in low-pressure systems, must be kept cool. Pressures in the 1×10^{-10}-torr range can be attained with butyl rubber O rings, provided they are subject to heavy compression pressure exerted by the bolted flange and are, in addition, cooled to temperatures of approximately 45°F by cooling water passing through passages adjacent to the rings in the flanges. If the temperature can be reduced to approximately 0°F ± 10°, this material is useful for pressures down to as low as 1×10^{-11} torr. However, it can only be used at such low pressures if good refrigeration systems are available to maintain its temperature in the range indicated.

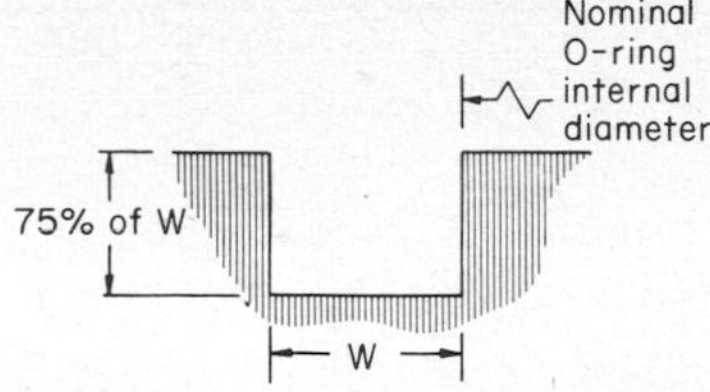

Fig. 14.1 O-ring groove dimensions for horizontal position.

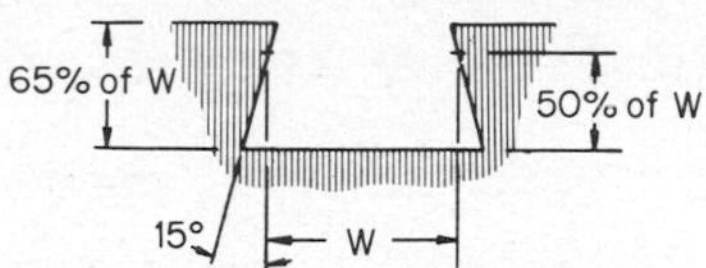

Fig. 14.2 O-ring groove dimensions for vertical position.

In connection with cooling systems for O rings, it should be borne in mind that if the rings are cooled too far, they freeze and cease to have elastic properties. Under these conditions, catastrophic leaks can and do occur. Usually, the place where this becomes critical is in a flange immediately adjacent to a liquid-nitrogen-cooled cold trap, which must be designed so that the temperature of the O ring does not go below approximately —10°F if this type of failure is to be avoided.

The groove in which the O ring is seated must be carefully maintained as to dimension and finished to at least a 32-rms finish. Various authorities give groove designs which yield different compression ratios to the O ring. Normally, these range from 10 to 25 percent. The writer's experience is that the rectangular groove shown in Fig. 14.1, which yields approximately 25 percent compression on the O ring, is highly satisfactory for horizontal flanges. It is highly important that the inside diameter (I.D.) of the groove coincide closely with the actual I.D. of the O ring, which must not be stretched to fit a groove too large for it, as failure of the seal always seems to result. For vertical O rings, it is sometimes desirable to use an undercut groove similar to that shown in Fig. 14.2. If desired, the outside of the groove may be left straight for easier machining with the undercut area on the inside of the groove. In this type of groove, the O ring is held firmly in place so that it has no tendency to drop out of the groove when the vessel is opened. This tendency can be very annoying when straight rectangular grooves are used on a vertical door.

In making up the O-ring closure, it is important that the groove surfaces be ground and polished such that no residual tool marks or scratches occur at right angles to the length of the groove. Such scratches, even though of a very minute character, will in all probability produce a leak which is difficult to locate and repair. Scratches parallel to the length of the groove do not, in general, produce leaks and can be tolerated provided their magnitude is small. In all cases, grooves should be ground and polished after normal machining operations are completed to produce as perfect a finish as possible. The mating flanges, which are flat and serve to compress the O ring, must have an equally good finish at the point where the O rings contact them. In the case of the flanges, transverse scratches are particularly troublesome. For this reason, a high degree of polish, produced by turning a part so that all polish marks are parallel to the length of the groove, is mandatory. Once properly prepared, O-ring grooves and mating flanges must be carefully protected during use or storage to prevent damage which could ruin all the effort which has gone into their preparation.

In making up the actual O-ring closure, the O rings to be used should first be cleaned to remove all dirt and foreign material, utilizing a *small* amount of acetone or alcohol, depending on ring material, on a Kleenex tissue as a cleaning agent. The O ring should then be given a very thin coat of vacuum grease applied by drawing it through fingers slightly coated with vacuum grease. We emphasize that the coat of vacuum grease should be *thin,* as considerable trouble can be experienced with seals where too much grease has been applied. The vacuum grease is not a sealing material but simply a lubricant to enable the O ring to distort its shape to fill completely the groove in which it is placed and compressed. The grease has no other function and cannot of itself provide any sort of vacuum seal. Much trouble has been occasioned by people who considered the vacuum grease to be a sealing agent and therefore applied it liberally. The result is just the opposite.

14.4 *Rectangular O Rings*

Closures in large doors of test vessels, furnaces, and the like are frequently too large to permit convenient use of circular O rings. For this reason, rectangular O rings are often used. As in circular O rings, however, these must be endless; and, if straight lengths of material must be used, they must be joined together and vulcanized and the joint later smoothed to a uniform surface before the O ring can be inserted in place. Open tapered joints in O rings, which are satisfactory for some pressure applications, cannot be tolerated for vacuum operation, and the O rings must be in one piece and endless. In all cases, the depth of the groove and the mating flanges should be such that when the flanges are brought tightly together by means of clamping bolts, approximately 25 percent compression will have taken place in the O ring. In most moderate-pressure vacuum-seal applications, clamping is necessary only to permit start of the pumping operation. After the system has been evacuated, the force of atmospheric pressure will serve to compress the joint; and, in the case of doors and the like for medium and high vacuum, clamps may be removed or loosened.

For ultrahigh vacuum, however, the situation is very different. Here, we must reduce permeation to the minimum possible amount; to do this, severe clamping pressures are required. It is therefore necessary to make use of bolted or clamped flanges to exert the necessary forces on the O ring. Where bolts are used, located outside of the O-ring circle, the designer must make sure that the thickness of the flange is sufficient so that the bending moments caused by the offset between the resistance of the O ring and the point of application of force from the bolts will not produce "coning" or distortion of the flange. This, if excessive, can result in the outer diameter of the flange coming metal-to-metal with the mating flange while still not exerting sufficient force on the O ring to compress it the required amount. If a C-type clamping device is used, the center of force and the center of resistance can be brought together so that this distortion of the flange does not occur. However, such clamping devices are somewhat awkward to use and expensive, and bolting is therefore frequently resorted to. As a result, flange thickness must, in general, be quite large, often on the order of 1 to 1½ inches in the case of 3-foot-diameter doors.

14.5 *O-ring Composite Gaskets*

The manufacture of flanges with grooves as described above is a time-consuming machining job. For this reason, it is sometimes desired to make use of preformed vacuum gaskets. These consist of a thin plate of aluminum or stainless steel to which are cemented O rings, usually two on each side. These can be installed in systems between two flat flanges without grooves; and when the flanges are pulled together to make contact with raised metal offsets outside the O rings, an excellent seal is possible. These are generally so made that it is possible to pump out the space between the two concentric O rings on either side of the diaphragm. The effect of this technique is to make use of the outer O ring for holding a relatively crude vacuum and subject the inner O ring to only a small differential between the high-vacuum side of its inner face and the partial vacuum on its outer face. As a result of this double-pumped system, the leakage per gasket can be greatly reduced. These types of seals are generally useful to pressures of approximately 1×10^{-7} or 1×10^{-8} torr, but not lower.

14.6 *Metal-sealed Flanges*

In ultrahigh-vacuum systems where baking is required no elastomeric O ring can be used in most cases. For such systems, metal-sealed flanges utilizing copper or gold wire are used on diameters of 10 inches

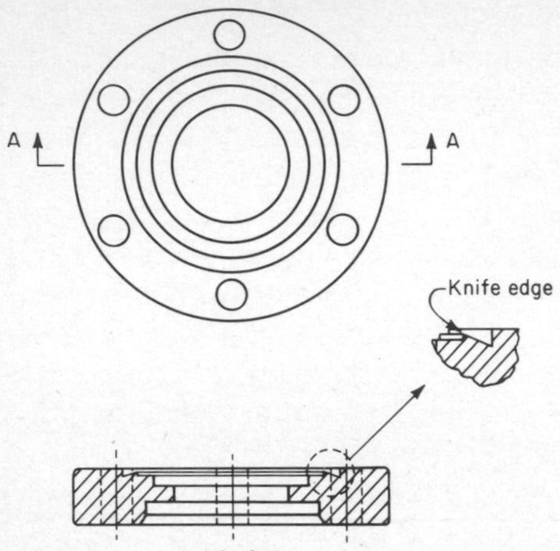

Fig. 14.3 Varian Conflat flange. (*Varian Vacuum Division, Palo Alto, Calif.*)

or larger; and copper flat gaskets, which are deformed by projections or knife edges in the mating flanges, are used for diameters below 10 inches. The finely machined flanges with the deforming knife blades machined and polished in their faces are available from several sources, together with matching bolts having approximately the same coefficient of expansion as the flange, which can be used for closure. The amount of compression is very carefully controlled by means of torque readings on the bolts, which are tightened by means of torque wrenches so that exactly the correct amount of pressure is applied. Such flanges will remain tight even when cycled repeatedly between 500°F and the temperature of liquid nitrogen. The leakage through properly made-up metal-sealed bakeable flanges is so low as to be undetectable by the best mass-spectrometer techniques available using helium tracer gas.

In general, the gasket material is used only once, being distorted by the first closure and afterwards thrown away and replaced at the next closure. In the case of copper, the reclamation value of the used wire or gasket is so low as to make reclamation impractical. In the case of gold wire, however, the reclamation value of the used gold is a very high percentage of the original cost of the wire; therefore, it is desirable to reclaim all such gold and have it reworked into new gasket material. Figure 14.3 shows the Varian "Conflat" flange with the details of the indenting knife edge. A similar knife edge occurs on the mating flange in order to produce the proper indentation of the copper gasket on both sides. For this flange, a flat copper ring is the gasket material.

For larger diameters, generally 10 inches and above, a copper wire formed gasket is used in place of the flat gasket. In these gaskets, a somewhat similar compression arrangement is used, in which the copper wire is confined within a small groove and indented by a knife-edge section under very high pressure. Techniques of this kind have been extended to diameters as large as 10 feet with good results.

The great problem in connection with all metal-sealed gaskets is to ensure perfect mating of the two flanges that are being forced together and proper torqueing on all the bolts that are used to pull them together. The use of torque wrenches to ensure proper pressures is essential, and the bolts themselves must be made of a material having as high a strength as possible combined with a coefficient of expansion comparable with that of the flanges which are being pulled together. If this is not done, the flange will open up after baking and cooling.

It should be emphasized that the elastomeric gasket technique is best limited to systems operating into the 1×10^{-8}-torr range, but not below. Below this value, metal-sealed gaskets are mandatory, and no shortcuts ought to be taken in an effort to avoid the expense of such gaskets. Repeated attempts to make use of less expensive closure methods in ultrahigh-vacuum systems have resulted in the universal conviction that there is no shortcut to tight closures in systems that must be baked.

14.7 *Bell-jar Seals*

In sealing a glass bell jar against the face plate, a situation arises where a flat gasket must be used. High compressive forces, which would permit the use of metal gaskets, are not possible, since in most cases the bell jar is fabricated of Pyrex glass. Several forms of gaskets are available for this purpose, but in general they all consist of neoprene, Teflon, or Viton-A formed into an angle shape which fits around the bottom of the bell jar and in service is clamped between the ground bottom surface of the jar and the machined and polished base plate. The force of the atmosphere on the bell furnishes the necessary clamping force. Figure 14.4 shows a typical gasket seal for a glass bell jar.

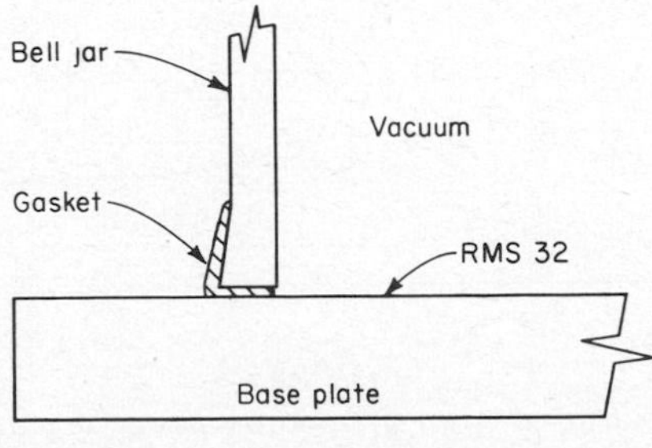

Fig. 14.4 Bell-jar gasket seal.

Such systems, utilizing neoprene or Teflon, are satisfactory for pressures to the low 1×10^{-6}- or high 1×10^{-7}-torr region. For systems which must operate at a lower pressure, metal bell jars are employed. With these, a flange is provided which permits clamping the bell jar to the base plate under adequate force to make a seal, either with organic elastomers for moderate pressures or with metal O rings for ultrahigh-vacuum pressures. However, see Sec. 14.3 for refrigeration techniques.

chapter 15

Accessories

15.1 *Introduction*

In any practical vacuum system, it is necessary to make use of a number of accessory pieces of apparatus for any of various reasons: to control the pumping operation; to permit the readout of electrical signals from inside the system during operation; to provide for the passage of electrical leads into the system; to provide for the passage of cooling water, air, or liquid nitrogen into the system for shrouds, cryopumps, or experiments being carried out in the system; to permit motions to be transmitted inside the system for manipulation of various experimental or production devices; and to provide for viewing the interior of the system through windows or to permit the introduction of simulated solar radiation. The nature of these devices must be sharply differentiated between systems operating in the medium- and high-vacuum field and those operating in the ultrahigh-vacuum field, due to the very much greater integrity requirements of the ultrahigh-vacuum system and the necessity of withstanding the bakeout temperatures associated with such systems. In what follows, we shall therefore attempt to distinguish between the appropriate items as applied to these two basic regimes of vacuum systems.

15.2 *Vacuum Valves*

Vacuum valves are required in almost all systems for use in the roughing line to isolate mechanical pumps from the chamber proper, where diffusion pumps form the fine-pumping portion of the pumping system. Such valves may be of either stainless steel or copper construction and may be applied either by O-ring flanges or by means of silver-soldering techniques, since the pressures in the roughing lines are not exceptionally low. The construction of the valve may be either of the rising-stem gate-valve type, the ball-valve type, or of the globe-valve type. They may be either mechanically or manually operated. In all cases, the ultimate seal is made by elastomeric O rings or disks which are forced against mating members to form the vacuum seal. For remotely operated systems or automated systems, the valves must of course be operated mechanically. The usual arrangement makes use of an air cylinder which through a suitable linkage actuates the stem, with an electrically operated pilot valve serving to turn the air pressure one way or the other as the result of signals from the control panel.

With the exception of the rising-stem gate valve and ball valve, all these types of valves are sealed around their stem by means of a bellows of stainless steel or brass, arranged so that no leakage path exists around the stem into the vacuum system. Such valves are entirely adequate for foreline purposes in all vacuum systems, since the minimum pressure they are expected to withstand is the blankoff pressure of the mechanical pumps, normally approximately 10 microns (1×10^{-2} torr).

The impedance drop across such valves is not, in general, of great importance, since the pressures in the foreline are within the viscous-flow region. Roughing line valves between the mechanical pumps and the chamber must be tight at the ultimate vacuum to be attained in the chamber. They must, therefore, be carefully chosen for use at this pressure level.

Valves for use between the diffusion pumps and the chamber fall into quite a different category. Here, impedance must be low in the molecular-flow region, and the valve opening should therefore be as large as possible, at least as large as the throat of the pump, and the flange-to-flange dimension of the valve should be as small as possible, again to reduce the impedance to a minimum value. These considerations dictate the use of thin or ultrathin models of gate valves between the diffusion pump and the chamber.

The most common types of such valves are made of cast-aluminum bodies impregnated with special sealants to render them vacuum-tight and employ a rising stem design similar to that shown in Fig. 15.1. An unfortunate condition of these valves arises from the fact that the stem

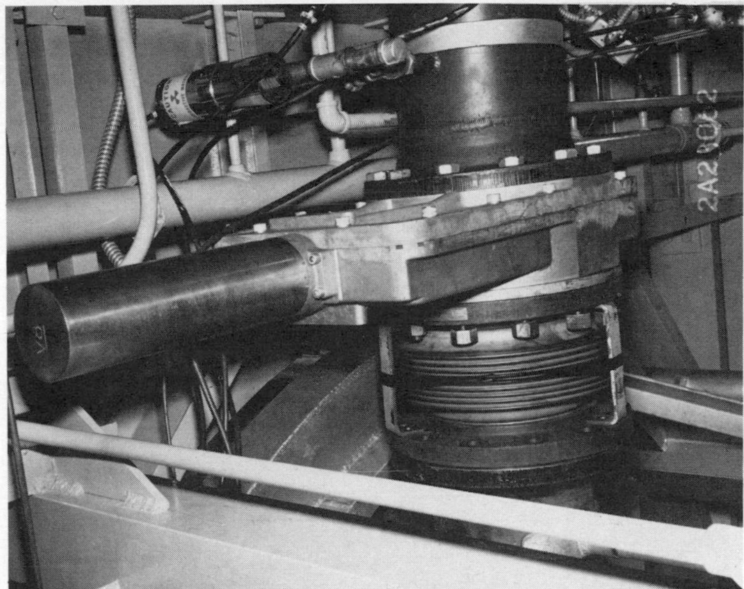

Fig. 15.1 Pneumatically operated vacuum gate valve.

must slide through a gland in order to open and close the valve plate, which is lowered to its correct location and then seated against the O rings by means of cam action of the traveling mechanism. As the stem moves through the seal, it carries into the vacuum chamber a certain amount of air which is occluded on the surface of the stem and which comes off as a gas burst as soon as the stem enters through the seal into the vacuum area. There is, in addition, a certain amount of leakage through the packing gland, regardless of the direction in which the stem is moving, which seems to be unavoidable during the period of motion. This difficulty could, of course, be avoided by the use of a bellows around the stem, but the large amount of travel necessary for large-aperture gate valves renders the bellows extremely long and rather fragile. For this reason, this construction is not often employed.

Another variety of gate valve has, however, recently come on the market. It employs a rotating shaft with internal cam motions to translate the rotation into a longitudinal motion of the gate. In this valve the seal around the rotating shaft consists of two separated chevron-type glands with a pumpout port provided between the glands. By evacuating the interglandular space to a roughing-pump pressure, the leakage can be held to very low values with the rotating seal, which in any case avoids the problems of the sliding shaft. Such valves have been made of ultrathin dimensions so that the longitudinal impedance due to the length of the valve and housing is reduced to a minimum.

None of the valves described above using elastomeric seats are usable in ultrahigh-vacuum systems where bakeout is to be employed. The only possible valve for this type of application is the metal-seat valve of the bakeable variety, as made by several manufacturers. In these valves, the ultimate seal is made by either a deforming nose piece or a deforming gasket of copper arranged so that it cooperates with hardened stainless steel surfaces and is wedged into place by a very high-torque driver. Most such valves are built with an internal torque-multiplying device and are seated by means of a torque wrench to a specified initial seating torque, which may be slightly increased on each successive closure until the limit of deformation of the mating parts is reached, at which time these deformable parts are replaced by fresh inserts and the process repeated. Such valves are capable of repeated bakeout without leakage in either the open or closed positions and have given excellent results in ultrahigh-vacuum systems. They are, however, quite expensive (approximately $375 for a 1½-inch size). For this reason, they are not widely used between diffusion pumps and systems in ultrahigh-vacuum systems.

On such systems it is customary to supply no valve between the final stage of pumping and the chamber proper, which makes it impossible to blank off the ultimate pump from the chamber for a quick shutdown. Instead, time must be allowed to cool down the final pumping unit below the boiling point of the oil before the chamber can be backfilled and opened. Rapid-cool coils on the pump base will speed up this process, which, however, still remains a fairly lengthy one. Most such systems employ a small backing diffusion pump which backs the large primary pump and is, in turn, backed by a mechanical pump. These backing pumps can, of course, be isolated by conventional elastomeric valves to permit individual checking.

The one valve that is necessary on any ultrahigh-vacuum system is the backfill valve. This must be of the bakeable type but can be of small diameter, since extremely rapid backfilling is not required.

On ion-pumped systems, no valve is employed between the ion pump and the system, but a valve must be employed between the system and the roughing pumps whether these are mechanical arrangements of blowers and mechanical pumps in stages or sorption roughing pumps employing molecular sieve material. In any case, the valve employed to isolate the roughing system from the chamber proper must be of the bakeable variety, since it will be exposed to full chamber bakeout and must hold the ultrahigh vacuum after roughing is completed.

15.3 *Expansion Joints*

Some sort of isolating joint must normally be employed between the mechanical pumps and the diffusion pumps attached to a chamber in order to absorb the vibration of the mechanical pump without introducing such vibration into the chamber proper. For systems operating at not lower than 1×10^{-5} torr, rubber sleeves made out of vacuum hose are frequently used between the diffusion pump and the mechanical pump. For lower-pressure systems, or those which are expected to operate for long periods, it is desirable to use metal bellows for this connecting joint. This will allow independent expansion of the various portions of the piping and will absorb the vibration from the mechanical pump.

Another reason for using metal bellows in the form of piping connectors is that for systems that must be leak-checked to extremely low values of leakage, it is necessary to eliminate rubber hose anywhere in the system since this hose is, in general, penetrable by helium. The leak detector therefore picks up not only helium from the suspected leak but helium from the air through the rubber hose connections. Thus it is necessary to use lengths of metal bellows for all connections where extremely low levels of leakage must be detected.

Metal bellows are available made of copper, brass, or stainless steel. For vacuum systems in general, the stainless steel bellows are preferable, since they can readily be welded to stainless steel flanges or other parts without difficulty, whereas the copper must be brazed, raising some difficulties.

Stainless steel bellows are available in two designs, both suitable for high-vacuum applications. The first of these is a simple convoluted tube made from a single piece of stainless steel. This is quite satisfactory for relatively small diameters but becomes rather stiff and inflexible when diameters exceed approximately 2 inches.

A more expensive type of bellows is formed by edge-welding a number of more or less "pie-plate"-shaped plates of stainless steel, as shown in Fig. 15.2, which shows a 10-inch bellows. This type of design is somewhat more flexible than convoluted metal, since the metal does not have to be as thick as in the case of the one-piece construction. It also lends itself more readily to repair should a small leak occur, since when it is removed from the line both the inside diameter and the outside diameter are accessible for rewelding. However, the cost is greater than the cost of formed bellows, and this type is probably not justified except for large diameters.

In general, any long run of large-diameter vacuum piping must include one or more bellows sections. These serve both to correct small misalignments due to fabrication errors and to provide for expansion or

Fig. 15.2 Ten-inch stainless steel bellows.

contraction due to temperature changes. If no such provision is made, incorrectable flange leakage usually occurs.

15.4 *Electrical Passthroughs*

In most vacuum systems, one or more electrical leads must pass through the chamber wall into the vacuum system and must do so without causing leakage of air into the system or degradation of the signals being handled. For medium-vacuum systems operating at pressures higher than 1×10^{-8} torr, seals of an elastomeric nature can be used surrounding the electrical leads. A wide variety of commercial passthroughs is available which answers these descriptions. Leads suitable for carrying currents ranging from milliamperes to thousands of amperes can be provided, and special ganged passthroughs are available in which as many as 100 leads can be carried through a 2-inch-diameter opening into the chamber, with cooperative plugs provided for both the inside and outside wire connections to the passthrough leads. Similar arrangements for coaxial leads are provided where high frequencies must be carried out of or into the chamber and where impedance effects from the mating metal parts would otherwise interfere with the signals being transmitted.

Special problems arise where thermocouple leads must come through the walls of the chambers, especially when cryogenic shrouds are being used inside the chambers. If the inside and outside of the passthroughs are at the same temperature, no effect will arise from a junction at this point provided the passthrough materials are similar in nature to the wires of the thermocouple being employed. Passthroughs are available with chromel-alumel, copper-constantan, and other thermocouple materials which, as mentioned above, allow proper reading of the thermocouples provided the temperature of the inside and the outside is the same. However, where cryogenic shrouds are in place inside the chamber close to the passthrough, the inner end of the passthrough pins will normally be at a temperature lower than the outer end of the same pin which is exposed to the room. Under these conditions, an emf is developed between the two junctions which will affect the readout of the thermocouple. At high temperatures, this effect is negligible, but at temperatures from 200°F down to —200°F and below, the effect of the junction at the passthrough is quite serious and, in general, this type of passthrough cannot be used. A hollow-pin-type passthrough can be used under these conditions provided that the thermocouple wire is run through the passthrough unbroken and further provided that it touches the hollow pin only at one point (the point at which the silver-solder closure is made) and is insulated from the pin for the remainder of the distance through the passthrough. An optional solution for this problem is to pass the thermocouple wire through a plate of plastic material, sealing it thereto by means of a suitable cement, and seating the plastic plate in a suitable flange with O rings. This construction permits proper readout of the thermocouple junction temperatures but introduces a nonbakeable passthrough into the system, which precludes the use of this arrangement on ultrahigh-vacuum systems. An alternate, and perhaps preferable, design is to locate the passthrough at the end of a long piece of tubing projecting far enough outward from the system so that its inside and outside remain at room temperature and the temperature effects noted above do not occur. This can be done at the expense of some additional work in preparing the passthrough arrangement.

For ultrahigh-vacuum systems where bakeout is necessary, passthroughs are normally made by means of ceramic insulators which are brazed into stainless steel flanges and which serve to insulate the passthroughs from the flanges by means of the ceramic. Such passthroughs

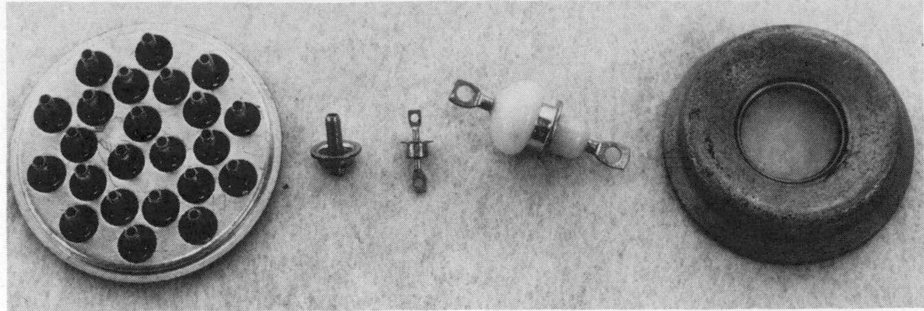

Fig. 15.3 Various types of passthrough fixtures.

are both expensive and more bulky than the types described above but do provide satisfactory performance with bakeable high-vacuum systems. They are available to cover the entire current range required, from 1,000 amperes to a few milliamperes, and have proved quite satisfactory in numerous systems. Figure 15.3 shows an assortment of bakeable passthroughs of a variety of types. Figure 15.4 shows large water-cooled high-current coaxial passthroughs of the nonbakeable type suitable for current of up to 1,000 amperes.

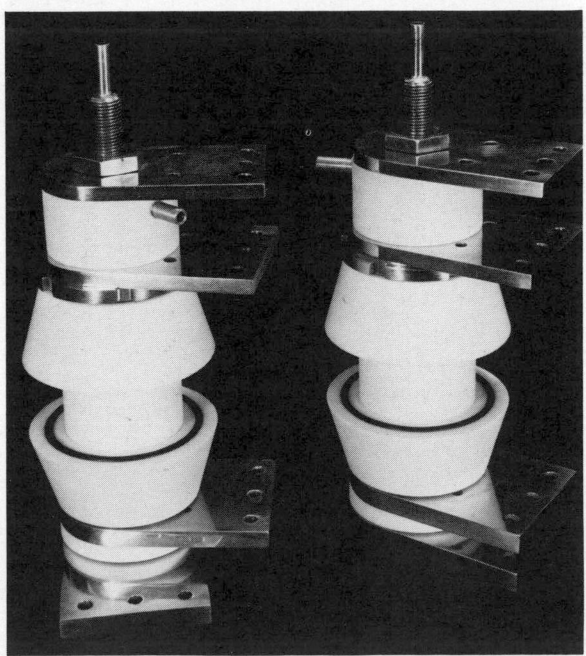

Fig. 15.4 High-current coaxial passthrough fixtures.

15.5 *Liquid and Cryogenic Passthroughs*

For passing water, compressed air, or cryogenic fluids to a vacuum chamber, passthroughs similar in design to the large electrical passthroughs are normally used. In the case of water and air no particular difficulties exist, and passthroughs of either the elastomeric or the ceramic-insulator type may readily be obtained. For cryogenic passthroughs, however, the problem becomes largely one of preventing undue heat flow from the vessel wall into the cryogenic tube carrying liquid nitrogen or low-temperature helium gas. Such passthroughs are generally of the bayonet type, in which a sleeve is welded to a stainless flange which is then attached in the normal manner to the chamber wall. This sleeve carries an inner tube arranged so as to be vacuum-jacketed by the internal pressure of the chamber, which connects through a special elastomeric seal with a bayonet-type tube coming from the source of cryogenic fluid.

The fluid is normally brought to the chamber in a vacuum-jacketed pipe terminating in the bayonet which slides within the chamber fitting, and a final seal is made between the inside of one pipe and the mating pipe by means of the elastomeric seal. This is not open to the vacuum, and any leakage that takes place can only represent gas loss out from the cryogenic tube to the outside. The clamps employed are normally sufficiently tight to reduce this loss to a minimum, since warm gas fills the inner space and acts as a blocking agent to prevent any further loss of the fluid. Such systems are used where detachability is required, as with internal shrouds in systems which must be removed periodically for cleaning or other maintenance. For permanent connections, as to cold traps and the like where no problem of disconnection is involved, simple vacuum-jacketed tubes are used, with the sealing membrane being drawn out into a long skirt to provide a long leakage path for heat from the chamber wall into the cryogen.

It should be reemphasized at this point that electrical leads inside vacuum systems must be insulated in a fashion appropriate to the pressure to be achieved. See Sec. 3.3 regarding wire insulation.

15.6 *Rotary and Sliding Glands*

Where actuating mechanisms must pass through the wall of a vessel, it is highly desirable to make use of rotating devices rather than reciprocating drives where this is possible. The most practical method, known as the Wilson seal, is shown in Fig. 15.5. This system, as developed by Wilson for use in the cyclotron, has since become very widely used

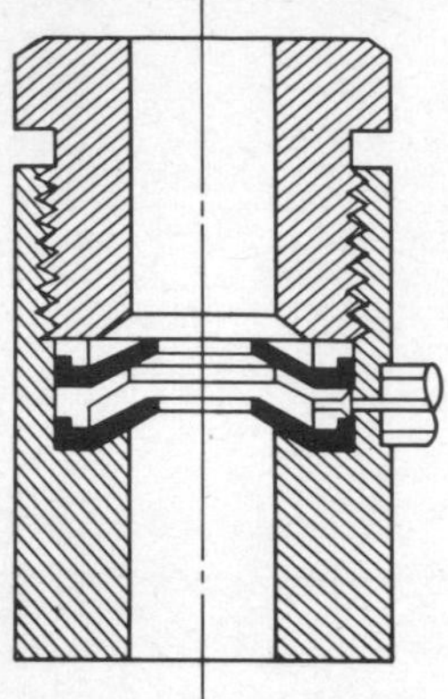

Fig. 15.5 Wilson seal.

in the vacuum industry, being available as a purchased accessory from most vacuum manufacturers. It consists basically of two deformed disk-type gaskets which are compressed so as to bear effectively against the shaft. The space between the two gaskets is evacuated by means of a pump connection to minimize leakage across the inner high-vacuum sealing gasket. Such systems have proved quite satisfactory for a variety of vacuum seals in the medium- and high-vacuum ranges. They can be used with quite good results on rotating shafts and with somewhat less success on sliding or reciprocating shafts.

The problem with any type of sliding seal is that the motion of the shaft through the sealing gaskets tends to produce longitudinal grooves on the shaft which serve as leakage channels through the sealing medium. It is therefore important, where reciprocating motion is required, to provide a hardened, highly polished shaft, very carefully guided in order to prevent scratching on the shaft surfaces. Even so, some leakage will generally occur in reciprocating shafts and will get worse with time. In addition to the leakage through the seal, there remains the problem of air carried in as occluded gas on the surface of the shaft. This air is fed into the system each time the shaft moves from the outside to the inside. This cannot be completely avoided, and it is therefore undesirable to use reciprocating seals in any sophisticated vacuum system.

A modified version of the Wilson seal is shown in Fig. 15.6. This employs alternate layers of Teflon separated by polished metal washers which are held in place by means of a thrust collar. In this seal, the closure is accomplished by flow of the Teflon under pressure from the thrust collar and requires, of course, a highly polished shaft. The advantage of this type of seal is that the low coefficient of friction of Teflon allows the device to function without the need of vacuum grease for lubrication, which allows achievement of a slightly better control of high-vacuum leakage than the plain Wilson seal with vacuum grease.

For ultrahigh-vacuum systems, none of these passthrough methods can be employed, since the leakage through any or all of these is too great to be tolerated and, in addition, the systems would not stand the bakeout temperatures. For ultrahigh-vacuum systems, therefore, all-metal transitions are used which, in the case of the reciprocating shaft, become a relatively long and flexible bellows fixed in a gas-tight manner to the shaft in its outer extremity and to the chamber wall at the flange. Guides must be provided for the shaft inside the vacuum

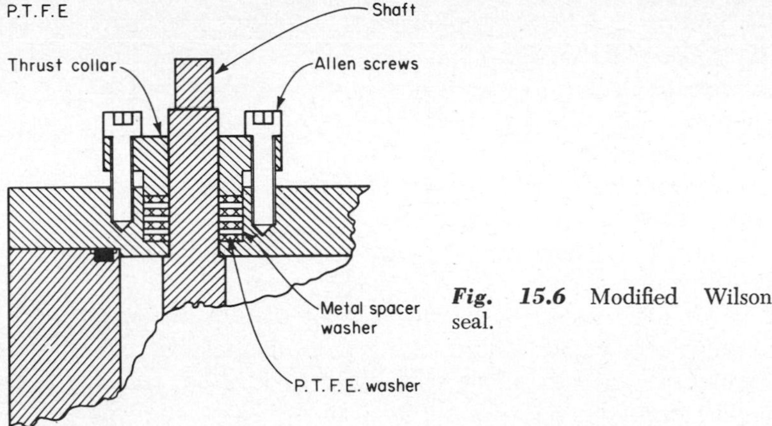

Fig. 15.6 Modified Wilson seal.

system as well as outside in order to prevent flexing the bellows other than axially, which would probably result in early bellows failure. Such systems are tight, but have limitations as to the amount of movement that can be achieved.

For rotary motion, the magnetic coupling represents the most suitable type of device to use with the ultrahigh-vacuum systems. This device makes use of a driven shaft to which is attached a series of magnets which cooperate with magnets or magnetic material placed on the inner shaft on the other side of a thin stainless steel plate. Rotation of the outer shaft causes similar rotation of the inner shaft, provided the torque is not greater than can be sustained by the magnetic field of the magnets. The inner shaft must, of course, be carried by bearings located inside the chamber, and the whole assemblage must be capable of being baked out. Generally, magnetic passthroughs of this kind will permit bakeout at 250°C but not at very much higher temperatures, since the magnetic field is deteriorated by too much temperature. Such magnetic passthroughs can be built to transmit reasonable amounts of torque, ranging from a few inch-pounds to 10 or 20 foot-pounds, at speeds ranging from a few to as many as 1,000 rpm. There is, of course, a speed limitation due to the fact that the flux is continually changing through the intervening stainless steel plate; and if the frequency is too high, eddy currents set up in the plate cause excessive heating as well as reducing the effective torque of the various magnets. However, this type of passthrough has been used in many bearing test apparatuses and operates at pressures of 1×10^{-11} torr and lower with good success.

15.7 *Liquid-nitrogen-level Controls*

As mentioned previously, cold traps are generally cooled by liquid nitrogen supplied from a dewar or vacuum-jacketed portable tank, known as an ARL bottle. It is necessary, in order to assure proper operation of the traps, to keep the liquid-nitrogen level within the traps at a reasonably constant value to assure proper cooling of the trapping surfaces. This may be done manually by the operator observing the liquid-nitrogen level and pouring additional fluid from a dewar as needed. However, for large systems or for those that operate for long periods, this can become extremely laborious and cumbersome; it is therefore desirable to make use of some automatic filling device.

Liquid-nitrogen-level controls are available from a number of suppliers, employing a sensing head which is alternately immersed in or just above the level of the liquid nitrogen in the cold-trap reservoir. This is connected through a control system to a valve connected to the source of supply of the liquid nitrogen in such a manner that when the valve is open liquid nitrogen will flow through the connecting line into the cold trap. In order to prevent undue loss of liquid nitrogen being transferred, the connecting lines should be vacuum-jacketed and the uninsulated connection sections kept as short as possible. For temporary installations, a plastic jacketing material is available that does a reasonably good job of insulating the transfer line. Such systems will operate unattended to keep the cold-trap liquid-nitrogen level at the proper height and are highly desirable for most systems. Several types of controls, ranging from carbon resistors to germanium diodes and even photoelectric sensors, have been developed which operate satisfactorily. If possible, it is desirable to adjust such a controller to operate intermittently on a long cycle in order to minimize line cool-down losses.

15.8 *Bakeout Systems*

We have several times referred to the necessity of providing means for baking out ultrahigh-vacuum systems and indeed, so far as possible, any systems that operate at pressures below 1×10^{-6} torr, where the type of O rings used in doors and flanges permits bakeout.

A number of different bakeout arrangements have been used successfully, all of which have certain advantages. The most common of these is the heating mantle, which is simply an insulating blanket with heating elements fastened to the inside surface thereof in a flexible manner so that the entire blanket can be placed around the chamber to be baked and held in place by straps or hooks of one kind or another. Such systems are relatively inexpensive and quite suitable for small vacuum chambers, but do have certain disadvantages:

1. They must be tailor-made to fit the particular system being baked, since holes must be provided for all flanges and passthroughs which protrude from the chamber body. They must further be insulated in

such a way that the heating elements do not come into actual contact with the chamber, which would of course short the elements.

2. Because of the necessary tailoring of the heating mantle to the system, it becomes quite difficult to introduce additional passthroughs, mass-spectrometer tubes, gauges, or other devices which may be desired at some date after the system has first been assembled. To make changes in the mantle usually requires a rather awkward reworking of the heating elements, since new holes cannot be placed in such a fashion as to ground out the heating elements.

3. The mantle normally must be removed from the system after baking, since it would retard the cooling cycle entirely too much if left in place. The removal of the hot blanket is apt to result in some tearing or damage to the fibrous insulating material of which the blanket is made, so that after repeated bakings, removals, and replacements the mantle eventually becomes so badly tattered that its replacement is necessary or elaborate repairs are required. For this reason, the mantle type of heating device has had a rather limited application, except in systems which do not require rapid cooling from the bake temperature and where the mantle can therefore be allowed to remain in place.

A second type of heating arrangement involves attaching heating elements to the actual vessel by bolting or welding in such a way as to permit ready heat transfer from the heating elements into the vessel. The elements themselves can be either tubular heating elements or flat air-heater-type elements, the only requirement being that they make reasonably good contact with the vessel wall to transmit the heat from the element into the vessel. Insulation in this case is usually provided by insulated, removable metal covers which bolt around the vessel to retain the heat during the bakeout cycle and which are removed at the end of the cycle in order to promote cooling. Such covers are usually fabricated of aluminum or stainless steel plates in order to promote good appearance in the complete chamber.

A variation of this method involves the use of permanently mounted covers spaced away from the chamber walls by an air space through which air can be circulated to promote cooling without removing the insulation. In the case of permanently mounted covers, a removable panel is usually provided over one section of the chamber shell to provide access to passthrough ports, either those installed in the beginning or to be installed at some future date. This section is usually made small enough so that no heaters need be installed on it, heat being obtained from adjacent heaters located on either side of the removable panel. In this fashion, access can be had to numerous passthroughs, and no great harm results from leaving off the covers over the passthroughs if the nature of the instrumentation being employed makes this necessary. Figure 15.7 shows an ion-pumped system with this type of bakeout system.

A third variety of heating device consists of a small furnace of the recirculated-air variety which can be lowered or wheeled into position to cover the chamber completely and which heats by circulating hot air around the chamber and back to heaters and a fan for recirculation. Figure 15.8 shows a diffusion-pumped system with this type of bakeout furnace.

In all cases, actual control of temperature is achieved by means of a contractor working in response to a pyrometer of the millivolt variety or potentiometer type or by a thermostat which can be set to maintain the desired temperature. In many cases, a time clock is provided, arranged in such a way that the heating cycle can be terminated after a set number of hours of baking without requiring the presence of an operator. This permits starting of bakeout at the end of the day's shift and terminating the bakeout some time during the night, thus having the chamber relatively cool when the day shift comes on again the following morning. Water cooling coils can be applied to the chamber to accelerate cooling after the termination of the bakeout, but the gain from the use of such coils is relatively small, since they can only be employed after the temperature has dropped to approximately 200°F, as otherwise steam generated in the coils can cause difficulty.

Fig. 15.7 Ion-pumped vacuum system with permanent backout covers.

Fig. 15.8 Diffusion-pumped vacuum system with removable furnace-type bakeout oven.

None of the bakeout systems described above are extremely complicated, all being sufficient, however, to achieve a reasonable uniformity of temperature over the vessel walls, doors, and other parts. Where bakeout is used, it is of course mandatory that all passthroughs, doors, etc., be sealed by metal gaskets in order to withstand the bakeout temperatures without damage.

15.9 *Windows*

It is frequently necessary to provide some form of viewing window by which experiments inside the vacuum chamber can be observed during operation. Small view ports are normally made of Pyrex glass sealed by glass welding techniques to a transition piece of Kovar or other metal sleeving which has a coefficient of expansion very close to that of Pyrex glass. This sleeve is then welded to a metal flange of stainless steel, which is then bolted to the chamber in the usual fashion. Windows of this variety are available up to 4 inches in diameter as stock items and up to approximately 6 inches in diameter, as special-order items.

Where solar simulators are used and a beam of simulated sunlight passes through a window into the chamber for the purpose of irradiating a specimen as it would be in space, Pyrex glass generally is not satisfactory. Unfortunately, all ordinary glasses have very poor transmission characteristics for light in the ultraviolet region, and the use of such windows therefore cuts off ultraviolet light in the beam from the solar simulator. By far the best material for transmitting the entire solar spectrum is clear quartz. This material transmits quite well, so far as light is concerned; but, having a very low coefficient of expansion as well as relatively low strength, it presents difficulties in making the joint with the chamber wall.

Satisfactory quartz windows, which will withstand moderate bakeout as well as ultrahigh-vacuum conditions, have been made in sizes up to 4 inches in diameter by brazing or welding the quartz to a special sleeve, generally of a noble metal, which is then welded to a stainless steel flange. Since the sleeve is relatively weak, a mechanical support is provided for the quartz to prevent overstressing the noble-metal sleeve.

Sizes of larger than 6 inches have not so far been successfully produced by the welding technique. However, windows of excellent quality up to 12 inches in diameter have been produced by RCA through the use of a ramming technique, in which a specially edged disk of quartz is forced into place in an alloy sleeve surfaced on the inner side with silver or nickel. When properly made, these windows have proved very satisfactory during normal usage.

A somewhat more rugged window results from the ramming technique as carried out by RCA, utilizing a sapphire disk instead of quartz. The advantage in using this material is that the strength of sapphire is considerably greater than that of quartz, and a more rugged structure therefore results. Unfortunately, the light transmissivity in the ultraviolet region of sapphire is not as good as that of quartz, so that some loss in ultraviolet transmission results. At the present date, the largest sapphire disks available are approximately 5½ inches in diameter, so that larger windows cannot be made out of sapphire. When still larger disks become available, however, this technique could undoubtedly be extended to larger windows as these become necessary. Full bakeout can be achieved using sapphire windows, since the strength is sufficient to withstand the stresses induced.

Windows intended for merely viewing the interior of the chamber can be made of ordinary plate-grade Pyrex glass with no special optical finishing. However, if the window is to be used for photography or for some types of readout devices of an optical nature, it is necessary that the windows be finished optically flat and parallel on both surfaces to a reasonable close limit. Flatness to two waves of sodium light is a reasonable specification for such windows, although closer tolerances can be provided if necessary. Quartz or optical-glass lenses can also be incorporated into chamber walls by these techniques, being used primarily for solar-simulator projection optics.

chapter 16

Finishing, Cleaning, and Backfilling Systems

16.1 *Introduction*

If a system is to pump down to low pressures, it cannot contain any material with a vapor pressure at the operating temperature that is higher than the operating pressure. If more than one contaminant is present, which is the usual case, the sum of the vapor pressures of each will be the limiting pressure in the system. Where the pump system can remove this, the system will eventually pump down, although the time may be long. The usual and principal offender in this regard is water vapor, which quickly contaminates mechanical-pump oil. However, other hydrocarbon vapors can be as bad or even worse, as they can not only contaminate the oil but frequently react chemically with it, increasing the difficulties.

Because of these considerations, it is extremely important that vacuum systems be very well cleaned prior to starting the pumpdown. Careful cleaning will always reduce the pumpdown time and may frequently permit the achievement of a lower ultimate pressure than could otherwise be attained.

16.2 *Mechanical Cleaning by Machine Methods*

We have emphasized several times the necessity for the interior of all vacuum vessels to be carefully ground and polished. The importance of this procedure is that it is quite impossible to properly clean parts that are oxidized, that have an as-cast surface, or, in general, have surface irregularities which can hold contaminants, including water vapor and oxygen. Such films, surfaces, etc., must be removed by mechanically machining and/or grinding all surfaces exposed to hard vacuum, followed by polishing, at least with a wire brush. Such treatment removes those films of oxide which otherwise would hold on to gases. However, the procedures outlined above will still leave the surfaces rough, with a tremendous surface area compared to the obvious internal area. The ratio between the true surface and the apparent surface can be reduced if the interior surface is polished to a mirror finish by means of a buffing wheel.

This may sound like a great deal of work, and indeed it is. However, if it is carried out in the early stages of fabrication, it is a great deal less work than if it must be done after the system is completely assembled. Wherever possible, plate to be formed into a vacuum vessel should be ground and polished in the flat before it is rolled up into circular parts. If this is done, final finishing will consist merely of internal grinding of weld surfaces and final touchup work, which is much less laborious than doing the rough finishing on the interior of a cramped or confined vessel.

16.3 *Hand and Chemical Cleaning*

At the end of the mechanical polishing operation, the interior surface will contain considerable quantities of by-products of the binders used in the buffing wheels, waxes and greases, and other materials coming from the mechanical polishing of the surface. These must be removed. For parts that permit it, a pickling operation is indicated, with the time so limited that no etching of the polished surface results. If this is not practical, then a scrubbing operation with Oakite or a good detergent in water is clearly indicated to remove waxes, oils, and greases so far as possible. After drying from the water-cleaning operation, which of course should be followed by a hot-water rinse, vessels which are small enough to permit it should be cleaned in a trichloroethylene-vapor degreasing bath to remove as far as possible additional grease that might be left over.

After this process is completed, we are ready to begin hand cleaning of the vessel. This should be done with acetone or alcohol, the latter being preferable, applied with "elbow grease" and care. In cleaning, the use of Kleenex tissues is much to be preferred over any other cleaning materials available. Rags of the usual machine-shop variety should never be used, since they load up with grease and oil and retain some of this even after laundering, together with residues of the detergents used in the laundering operation. Invariably, they leave about as much contamination as they remove.

Kleenex tissues are pure cellulose and are not loaded with cold cream, perfume material, or fillers as are certain other tissues intended for facial use. The Kleenex tissue should be wet with acetone or alcohol from a squirt container and then used as a swab to clean the surface until the tissue becomes discolored. It should then be immediately discarded and a fresh tissue used. Repeated cleanings of this kind, until an acetone or alcohol wipe results in no discoloration of the tissue, should be carried out to clean all contaminants from the interior surfaces of the vacuum chamber.

Warning: Acetone or alcohol vapor is toxic and it is necessary to make use of ventilating fans directing a current of fresh air toward the man cleaning a vacuum vessel. The vapors are also explosive when mixed with air, and the ventilating system in the room where the cleaning is being carried out must be in excellent working order to dilute these vapors and expel them from the building. In all cases, a ventilating fan should be directed into a chamber where vapors might accumulate in a confined space, thus giving the operator all the symptoms of a cheap drunk as well as creating a serious fire hazard.

The discarded Kleenex tissues will still contain acetone or alcohol. They should immediately be placed in an approved step-on safety container. At the end of each shift, these containers should be emptied into a closed metal container located outside of the building and the contents of this container disposed of in the normal hazardous-material disposal method. Care must be exerted in this connection, since spontaneous combustion can take place in a closed vessel containing combustible materials soaked in either alcohol or acetone, and some safe method of disposal should be provided. Serious fires have resulted from failure to observe these precautions.

It goes without saying that any open fire, particularly smoking on the part of the operators, is forbidden under these conditions. Operators should preferably wear rubber gloves, since too long exposure to these cleaning materials produces dermatitis in sensitive individuals and skin irritation in practically everyone.

Where alcohol is used, it should preferably be ethyl alcohol, and any denaturant used should be one which will leave no residue within the chamber. Some of the sulfur-bearing denaturants produce more contamination than the alcohol can remove.

The above remarks regarding cleaning of the vacuum vessel apply with equal urgency to anything placed inside the vessel during operation. This includes electrical wiring, mechanical parts, gears, and plastic materials that may be used in various test configurations. These, so far as possible, should be thoroughly cleaned before being placed inside the chamber. It is necessary to use care in selecting the cleaning agent for various types of plastics, since some of them are soluble in some of the cleaning agents mentioned.

16.4 *Backfilling*

In backfilling vacuum vessels after a run is completed it is extremely important to make use of a dry gas for backfilling. The preferable agent for all diffusion-pumped systems is argon as evolved from a dewar of liquid argon, which is not only dry but inert and does not readily adhere to the surfaces of the chamber. However, a good substitute for argon is nitrogen gas obtained from a liquid-nitrogen-filled dewar, which has proved extremely satisfactory in many cases. Argon should not be used on ion-pumped systems because of the poor pumping speed of ion pumps for argon.

Where it is not convenient to use either of these relatively inert and very dry gases, air can be used provided the air is first dried of all moisture. If room air is used, the humidity of the day in question has a very noticeable effect on the difficulty of repumping the chamber on subsequent closures. In the case of a chamber operating at 1×10^{-10} torr, an additional time of 4 to 8 hours was required to reach pressure after exposure on a humid day as compared to a dry day. For this reason, it is usual to install such systems, where possible, in controlled low-humidity rooms, since the cost of dehumidification is more than repaid by the shorter cycles made possible thereby. In any event, the first letdown to atmospheric pressure should be done with a very dry gas, preferably an inert gas such as nitrogen or argon, as the first monolayers that are absorbed on the chamber walls are the most difficult to remove; and to some extent the use of a dry letdown gas will prevent picking up as much water vapor from the atmospheric air during the period when new work is being placed in the chamber as would otherwise occur. This is mainly true when the exposure time is quite limited. For long exposures, water pickup is inevitable.

When an experiment or production run being carried out in the chamber liberates large volumes of decomposition products, perhaps organic in nature, these will probably contaminate the chamber so that fairly complete cleaning with acetone or alcohol will be required before the next pumpdown. In some cases, the contamination may be so bad that a reconditioning bakeout will be required before the chamber can again be taken down to low pressures. However, no general rule can be made in this connection, since the effect depends critically on the nature of the materials being deposited on the walls and their vapor pressure and tendency to acquire monolayers of water vapor. In most cases, a good cleaning with acetone or alcohol will suffice to remove most of the wall deposits and allow a satisfactory repumpdown to be achieved.

chapter 17

Leak-detection Techniques

17.1 *Necessity for Leak Checking*

The necessity for accurate leak checking of vacuum systems is frequently misunderstood. It is assumed that a pressure test, perhaps with soap-bubble tracing at welded joints, can be substituted for more sophisticated methods of leak-checking the vacuum systems.

It is an accepted method in pressure vessels to seal all joints and then pressure-test the structure by applying several times the working pressure. If the vessel holds this pressure, as indicated by an ordinary gauge, overnight or for 24 hours, it is then assumed to be leak-free. Unfortunately such a vessel, indicating no leak under a test of this sort, may have an unworkably large leak when operated at low pressures. For example, if a system had a volume of 439.6 cubic feet and the leak rate was 1 cubic inch per hour and the pressure 1 micron (1×10^{-3} torr), it would then be necessary to pump out 439.6 cubic feet of gas at 1 micron pressure every hour in order to maintain the pressure. In a pressure vessel of the same capacity at 15 pounds pressure, the same leak rate would be only $\frac{1}{760{,}000}$ times the volume and in one day only $\frac{3}{95{,}000}$ times the volume

TABLE 17.1 Gas Constants

Name	Formed	Atomic radii $\times 10^{-8}$ cm	Molecular radii $\times 10^{-8}$ cm
Hydrogen	H	0.37	1.17
Helium	He	0.465	
Nitrogen	N	0.55	1.57
Fluorine	F	0.68	
Carbon	C	0.77	
Chlorine	Cl	0.97	
Argon	A	1.91	1.47
Carbon dioxide	CO_2	.....	1.61
Carbon monoxide	CO	.....	1.56
Freon 11	CCl_3F	.....	1.73
Freon 12	CCl_2F_2	.....	1.73
Freon 21	$CHCl_2F$	.....	1.73
Freon 22	$CHClF_2$	.....	1.73
Freon 114	$C_2Cl_2F_4$	.....	3.41

would escape. It would indeed take an extremely sensitive gauge to detect the pressure change of 0.003 pound per square inch in a system operating at 15 pounds. In other words, a system that was perfectly tight for the purposes of a pressure vessel would have a leak of catastrophic size when used as a vacuum vessel, due to the high multiplication factor inherent in vacuum operations.

It can be seen from Table 17.1 that the molecules with which we are dealing can penetrate extremely small holes. The units are 10^{-8} centimeters, and all the radii are in the direction of the smallest dimension of the molecule, which of course gives us the most trouble. Because of the small atomic and molecular radii of the elements, it is apparent that extremely small holes can yield appreciable leak rates. In the case of most of the elements, the molecular radii are the important ones. In the case of a few of the elements where diatomic molecules do not form, the atomic form determines the permeability.

Because the dimensions of the atoms penetrating the leak are so very small, leaks no larger than those caused by slight porosity in welds can cause serious trouble in ultrahigh-vacuum systems.

17.2 *Leak Checking by Means of Acetone and Gauges*

Many vacuum systems as originally assembled have in them so many leaks that they cannot be evacuated to a point permitting the helium-leak detector to be used. Under these conditions, leaks can be located more simply. If a thermocouple gauge is connected on the pump side of the system such that it reads manifold pressure and is evacuated to a range at least below 200 microns (2×10^{-1} torr) and if at this point a small amount of acetone is sprayed or painted on a spot where a leak exists, an immediate pressure rise will occur in the gauge, which will return to its former reading as the acetone evaporates and is expelled from the system.

This technique of testing by means of a thermocouple-type vacuum gauge is sufficiently sensitive to allow the location and elimination of all of the larger leaks in the system. These must generally be located and repaired by welding, or other repair of a permanent nature, before the system can be further evacuated.

It is important that repair not be accomplished by the use of Glyptal lacquer, wax, or similar temporary repair agent, since in general these materials are hard to remove and render repair by welding all but impossible. It is therefore better to locate the major leaks by the acetone technique and to reweld these leaks before proceeding further with the testing. Completion of the acetone checks will generally enable the system to be evacuated into the range where a mass-spectrometer instrument can be used for checking the smaller leaks.

17.3 *Helium-leak Detectors*

Helium-leak detectors are basically mass spectrometers of a type where the readout is confined to the helium line (see Fig. 17.1). They consist essentially of a vacuum pumping system with gauges and a mass-spectrometer tube where the entrained gases are ionized by electron bombardment and then accelerated to a negative collector. A magnetic field causes the accelerated particles to spread into a spectrum depending upon their mass, so that their various components can be isolated. A receiver collects the ions which fall at the helium location and, after passing through a suitable amplifying circuit, shows a reading on an indicating dial. Sensitivity controls are provided which enable the instrument to work over a variety of sensitivity ranges, the minimum being approximately 1×10^{-11} atmospheric cubic centimeters per second unless an electron multiplier is used, in which case a sensitivity of 1×10^{-14} atmospheric cubic centimeters per second is possible.

The sensing arrangement of these instruments can be used in two ways. One makes use of a small suction probe with a restricting orifice in order to allow the vacuum pump necessary for the operation of the mass spectrometer to work; the other consists of a direct connection with the vacuum system being checked. The probe technique is used

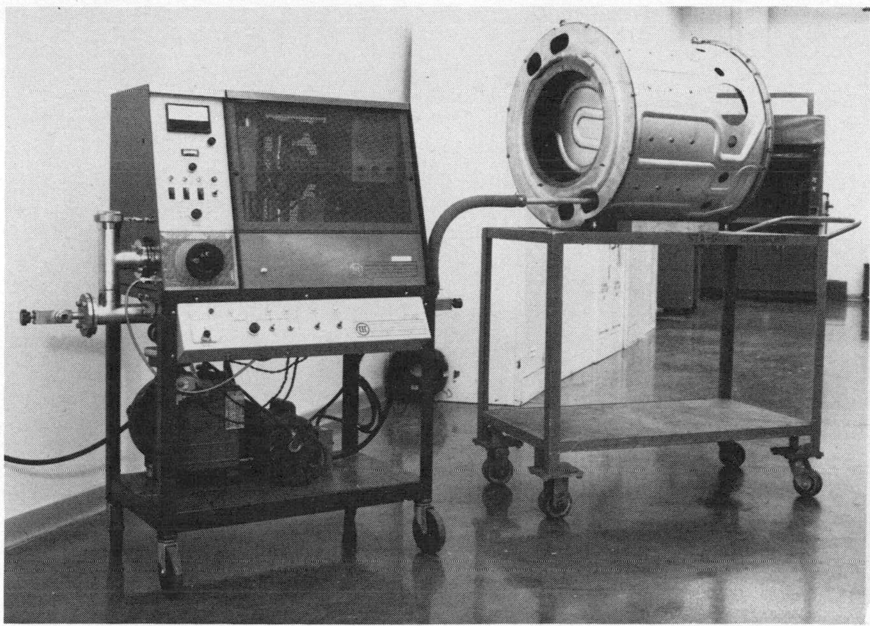

Fig. 17.1 Helium-leak detector shown leak-checking a cryogenic shroud. (*Consolidated Electrodynamics.*)

when a vessel is pressurized with helium and the sniffer applied at suspected outside points to detect helium leaking out into the atmosphere. Unfortunately, this method lacks extreme sensitivity, since the orifice necessary in the sniffing probe prevents much inward motion of air or helium molecules and hence cannot detect extremely small concentrations. The chamber can be operated at only slightly above atmospheric pressure using this probe.

If the leak detector is connected directly into the vacuum chamber, it is necessary that the pressure in the chamber be reduced to approximately 1 micron (1×10^{-3} torr) before the leak detector will operate. Under these conditions, however, the detector is extremely sensitive and can detect leaks of small size. It must therefore be so used for any high-vacuum system. When the leaks are of very small size, the leak detector is usually connected to the foreline of the diffusion pump to permit its use at system pressures too low for it to be in direct connection.

Operation in either case consists of squirting the outside of suspected areas with a small blast of helium from a pressure tank in such a way that if there is indeed a leak at that point the helium will pass through it into the vacuum system and be intercepted by the leak detector. It is possible to check total leakage by enclosing the vessel to be checked in a plastic balloon or bag filled with helium. In this case, the total leakage of all joints will add together; and the instrument, having been calibrated with standard leaks that are furnished, can be used to measure the flow of helium through the total of all the leaks into the system. Unfortunately, this technique does not tell where the leaks are, and it is necessary to go to the probe application of helium in order to locate leaks and thereby permit their repair. When helium-leak checking one starts checking at the bottom of the pressurized system and at the top of the evacuated system. The reason is obvious when one remembers that helium is considerably lighter than air.

When checking for leaks with any of these systems and attempting to locate the leaks, it is important to remember that a checking gas will drift from the place where it is applied to other parts of the system and can easily give false indications. It is therefore *highly desirable* when using the evacuated method of helium-leak checking to employ a fan blowing away from the system, not directly on the helium nozzle. When this is done, the actual leak is easier to locate and thus repair. Helium will also penetrate readily through rubber vacuum hose used to connect the leak detector or between the mechanical and diffusion pumps. Metal bellows-type hose or solid-pipe connections must therefore be used when maximum sensitivity is required.

Leak checking is an extremely slow procedure because it is usually necessary to correct any larger leaks found before smaller leaks can be located. The use of vacuum putty, dum-dum, or duct-seal may be resorted to in order to close one leak in a temporary manner so as to permit checking other suspected leaks. When several such leaks are so covered, be aware that helium can slowly penetrate the vacuum putty or dum-dum so that if too many such temporary closures are made a constant helium signal may be presented by the leak detector, raising the background to a point where small leaks become invisible. It is better to stop and repair by welding all leaks discovered, and then resume testing.

When dum-dum has been used, extreme care must be used to clean all traces of the material from the location where welding is to be carried out. Failure to do so will cause porous and leaky welds, which will require repair the second time around.

This makes it necessary to carry out the process in several steps alternating between checking, rewelding or reworking, checking again, reworking again, etc. It is also unfortunately true that vacuum systems which are shipped across the country, though tight before shipment, are seldom tight upon arrival. Usually all joints will have to be retightened and the system rechecked before integrity can be reachieved after movement. This is, of course, somewhat less true in the case of small systems than in the case of large ones.

After all major leaks in a system have been located, it may become apparent that small leaks exist around suspect penetrations, flanges, or other areas which are difficult to locate sufficiently accurately to permit repair. Such leaks may be localized by masking off the suspected areas with plastic sheeting and Scotch tape, then injecting helium into the masked-off area only by means of a hypodermic-needled type of probe termination. By masking only a very small area at a time (and using only a small amount of helium gas) the actual leak can usually be located. Warning: The location of one small leak around a flange or passthrough by this technique does not preclude the existence of another small leak near or opposite the one discovered. Such leaks often occur in pairs, since the poor technique which caused the one first located probably continued during all the welding operations on the area in question and may have generously contributed additional leaks.

17.4 *Repairing Leaks*

Small leaks in welded parts can usually be repaired by grinding out and rewelding the part. However, if the repair is not successful on the first rework, it may be necessary to scrap the offending part or to cut out the metal adjacent to the joint and patch with a new and larger piece. The act of welding a joint affects the adjacent metal by changing its grain structure and in some cases by changing its composition. The effect is to create porosity and strength loss in the heat-affected zone such that leakage persists even after the most careful reworking. This is noticeable in the case of stainless steel, but much more so in the case of aluminum, where rewelding usually requires cutting out the entire old weld section and substituting a new piece which may be rewelded on entirely new material.

The reader is referred to the American Vacuum Society Tentative Standard on "Helium Mass Spectrometer Leak Detector Calibration" for methods of calibrating leak detectors. Such methods give a means of leak-detector calibration which ought to be carried out at frequent intervals, since sensitivity normally deteriorates with time, being seriously off after only a month's average usage.

17.5 *Troubleshooting*

After a system has once been checked out and found to be tight to the limits required, it frequently develops leakage again at a later date. The first place to check is obviously at all penetrations of the shell, whether for water, oil, electrical leads, or mechanical joints. This definitely includes threaded joints where gauges and pipes are fastened into the vacuum system, particularly if these are sealed in part by Glyptal lacquer. If all of these joints turn out to be above suspicion, one should then check again around welded areas. We have found frequent instances where water-cooled welds develop porosity after some months of operation, not at the weld proper but adjacent thereto, requiring reworking after some initial period of satisfactory operation. This effect may be due to slow corrosion in areas of the weld where some contamination exists, permitting the corrosion to eat through the stainless steel, producing a microleak. This, of course, must be repaired by grinding out and rewelding.

17.6 *Total-leakage Measurements*

In order to determine the net pumping speed available for handling gas loads from work to be carried out within the chamber, a total-leakage-rate determination is often made of the empty chamber. To accomplish this, the leak detector is connected to the foreline of the diffusion pump. If the chamber employs more than one pump, all but one are blanked off during the net-pumping-speed run. The chamber is now pumped down as far as possible with the leak detector running but blanked off from the system. The leak detector is now opened into the foreline, and when stability has been achieved the mechanical backing pump is slowly valved off. If the leaks are sufficiently small, the leak detector will back the entire chamber without difficulty. This method has been successfully employed even where the diffusion pump was of the 48-inch size.

The leak detector as attached to the chamber must now be calibrated. This is done by means of a known helium leak attached to the chamber by means of a good valve, at a point some distance from the pumping port. A reading of the leak detector with the helium leak closed is first taken. The leak is then opened and a second reading taken. The difference between these two readings represents the deflection caused by the known leak. The standard leak in terms of standard cubic centimeters per second is divided by the deflection in divisions of scale. The result is the leak-detector sensitivity, as connected to the chamber, in standard cubic centimeters per division. Such a reading should be repeated several times, and the results averaged.

Typically, the leak-detector sensitivity as connected to the chamber will be less than the sensitivity of the leak detector measured with the standard leak when disconnected from the chamber and blanked off. Thus a leak detector might have a sensitivity blanked off of 1×10^{-10} standard cubic centimeters per second and a sensitivity as connected to a large system of 2×10^{-9} standard cubic centimeters per second.

To check the total leakage of the system, a polyethylene bag is now made up (from sheeting and masking tape) completely enclosing the system except for the leak detector. With the helium leak closed and all conditions well stabilized, the bag is now filled with helium. For large systems a helium-concentration meter is used to sample helium at the bottom of the bag. Usually a concentration of at least 50 percent at the bottom is required. Filling should be rapid; if possible, in not more than 10 seconds.

As the bag fills, the leak-detector deflection will increase, and measurements may require decade changing. Readings should be taken every second and recorded. The additional deflection above the base value before the helium was admitted, multiplied by the leak detector sensitivity in atmospheric cubic centimeters per division, will give the total system leakage in atmospheric cubic centimeters per second. To determine the pumping speed required to handle this amount of leakage at a given operating pressure P,

$$S = \frac{\text{At. cc/sec} \times 760}{1{,}000 \times P} \qquad \text{liters/sec}$$

The leak-detector response time, especially in large systems, will be appreciable. This is normally given by $t = V/S$. For the system described above, where the leak detector pump backs the diffusion pump completely, both V and S must be multiplied by the appropriate pressure; thus

$$t = \frac{VP_{\text{chamber}}}{SP_{\text{leak detector}}}$$

Response time in large systems will normally be from 1 to 3 seconds.

In systems having large openings sealed with elastomeric O rings, it must be remembered that these O rings are penetrable by helium and that they also absorb and hold helium, releasing it for some time after the helium has been released from the bag. After 3 seconds, helium penetration of the seals begins to show up in many cases, and after 6 seconds it becomes the dominant feature in the leak-detector reading.

Subtracting the pumping speed required to overcome the numerous small leaks, as obtained above, from the total available pumping speed, will give the net pumping speed available to overcome the expected outgassing of the work. If you are so unfortunate as to find the speed available at the required working pressure near zero, or even worse, a negative quantity, you are in good company. The only solution is to go back again to leak checking.

For ion-pumped systems, this type of testing must be carried out with the ion pump turned off and the system being roughed by the mechanical vacuum system. Such systems usually consist of a two-stage mechanical pump plus a Roots blower. In this case, the leak detector is used to back the Roots blower. Effective runs can be made provided the pressure can be reduced to the range of 1×10^{-5} torr with the ion pump turned off. To reduce outgassing sufficiently to accomplish this will usually require baking, then pumping the system to a pressure at least as low as 1×10^{-6} torr, then shutting off the ion pump and allowing its pumping rate to decay to near zero—which may require an hour or more—before beginning the test.

In the case of a large ion-pumped system, wall out-gassing—mainly of water vapor—may be so severe that the leak detector cannot back the blower. In such cases, the cooling of a liquid-nitrogen-cooled surface, either inside the chamber or in the form of a cold trap in the line leading from the chamber to the blower, is permissible. Such a cold surface will pump the water vapor but none of the other gases present. Since the water vapor comes only from the walls and interior parts of the chamber and not from the true leaks to any extent, its pumping other than in the leak detector does not at all invalidate the helium leakage being measured. Unless the leaks are very large, this procedure will allow the leak detector to back the Roots blower satisfactorily.

chapter 18

Theory of Gases

18.1 *Introduction*

In order to understand and use the equations which describe the flow of gases through orifices and piping, it is necessary to have a general understanding of the nature of the gas and of gases in general. With such an understanding, the proper formulas can be selected to compute the actual characteristics of a real system and thus predict the performance or determine the size of pumps, valves, and accessories that will be needed to produce a desired performance. More detailed treatments of these phenomena may be obtained from a number of useful books on vacuum or on physics, which will be found available in many places. There have been some recent advancements in an understanding of the performance of gases in vacuum systems which are summarized well by Van Atta and Dushman. In the following, we shall simplify the equations to the maximum degree possible while still retaining an understanding of the phenomena occurring in the vacuum systems.

18.2 *The States of Matter*

All matter exists in one of three states: solid, liquid, or gaseous. A solid may be defined as a body possessing both definite volume and definite shape at a given temperature and pressure. Under an applied force, a solid may be distorted or suffer a change in dimension; but, provided the elastic limit has not been exceeded, removal of the force will restore the body to its original condition. A liquid in bulk, on the other hand, has a definite volume but no definite shape, while a gas has neither definite shape nor volume. Liquids and gases are both termed fluids, and both offer no resistance to shape deformation. Inertia and viscosity may resist changes in shape, but these forces have no tendency to restore the fluid to its original form. A liquid, insofar as it fills the container, will always adopt the shape of the container in which it is placed but will retain its definite volume, while a gas will always fill completely any container in which it may be confined.

18.3 *The Nature of Gases*

A gas may be regarded as consisting of molecules traveling in straight lines at random and at high rates of speed within the containing space, and colliding frequently with other molecules or with the walls of the container. The force exerted per unit area on the walls of the container by the colliding molecules is known as the pressure—a force present at all times and distributed uniformly over the entire surface. The fact that small molecules produce a considerable bombarding force upon container walls suggests that the number of collisions with the walls must be large and that the molecules must be moving with high velocities.

The space occupied by the molecules themselves within a gaseous volume is a small fraction of the total volume of the gas under ordinary conditions of temperature and pressure. Thus if all the air in a room 20 feet by 10 feet by 10 feet were liquefied, the volume of the liquid would be approximately 2.4 cubic feet, or about 0.1 percent of the volume of the room, and yet the molecules would not be touching each other. Hence, we may conclude that molecules generally are separated from each other by distances which are large compared to molecular diameters, and that within a gas the space actually occupied by molecules is very small, most of the volume being "free" space. This accounts for the much lower densities of gases as compared to liquids and solids.

Also, this large amount of free space within a gas makes compression of the gas fairly easy. The compression process merely reduces the large free space and, by reducing the average distance between the molecules, brings them closer together. When there is no attraction between the molecules, the decrease in free space on compression is equal to the observed decrease in the total volume of the gas. Similarly, on expansion the average distance between molecules is increased, and thereby also the free space of the gas. In any case, the random motion of the molecules will give the effect of completely filling any containing vessel in which the gas is placed.

In terms of the structure of a gas outlined above, it is easy to understand why gases interdiffuse or mix. Two different gases, such as nitrogen and oxygen or any number of nonreactive gases, when placed in a container will by their motion mix with one another very quickly regardless of density. This mixture of gases will in many respects behave like a single gas, and the molecules of the various gases will collide with each other regardless of similarity or dissimilarity. Further, the total pressure exerted by the mixture will be determined by the total number of collisions between the molecules of all kinds and the walls of the container, a pressure to which each particular kind of molecule contributes its share.

18.4 *Boyle's Law*

In 1662, Robert Boyle reported to the Royal Society of England the results of his studies on the relation between the volume and pressure of a gas at constant pressure. Boyle confined within a graduated tube a quantity of gas and then measured the volume of that gas under different applied pressures. He found that the volume decreased with increasing pressure and that, within the limits of his experimental accuracy, the volume of any definite quantity of gas at constant temperature varied inversely as the pressure on the gas. This highly important generalization is known as Boyle's law. Expressed mathematically, this law states that at constant temperature $V = 1/P$ or that $V = C/P$, where V is the volume and P the pressure of the gas, while C is a proportionality constant whose value is dependent on the temperature, weight of gas, its nature, and the units in which P and V are expressed. On rearrangement, this equation becomes

$$PV = C \tag{18.1}$$

from which it follows that if in a certain state the pressure and volume of the gas are P_1 and V_1, while in another state they are P_2 and V_2, then at constant temperature $P_1V_1 = C = P_2V_2$

and

$$\frac{P_1}{P_2} = \frac{V_2}{V_1} \tag{18.2}$$

18.5 *The Charles or Gay-Lussac Law*

Charles observed in 1787 that the gases hydrogen, air, carbon dioxide, and oxygen expanded an equal amount upon being heated from 0 to

80°C at constant pressure. However, it was Gay-Lussac in 1802 who first made a quantitative study of the expansion of gases on heating. He found that for all gases the increase in volume for each degree centigrade rise in temperature was equal approximately to 1/273 of the volume of the gas at 0°C. A more precise value of this fraction is 1/273.16. If we designate by V_0 the volume of a gas at 0°C and by V its volume at any temperature t°C, then in terms of Gay-Lussac's finding V may be written as

$$\begin{aligned} V &= V_0 + \frac{t}{273.16} V_0 \\ &= V_0\left(1 + \frac{t}{273.16}\right) \\ &= V_0\left(\frac{273.16 + t}{273.16}\right) \end{aligned} \tag{18.3}$$

We may define now a new temperature scale T such that any temperature t on it will be given by $T = 273.16 + t$, and 0°C by $T_0 = 273.16$. Then Eq. (18.3) becomes simply

$$\frac{V}{V_0} = \frac{T}{T_0}$$

or generally

$$\frac{V_2}{V_1} = \frac{T_2}{T_1} \tag{18.4}$$

This new temperature scale, designated as the absolute or Kelvin scale of temperature, is of fundamental importance in all science. In terms of this temperature scale, Eq. (18.4) tells us that the volume of a definite quantity of gas at constant pressure is directly proportional to the absolute temperature, or that

$$V = kT \tag{18.5}$$

where k is a proportionality constant determined by the pressure, the nature and amount of gas, and the units of V. The above statement and Eq. (18.5) are expressions of Charles' or Gay-Lussac's law of volumes.

18.6 *The Combined Gas Law*

The two laws already discussed give the separate variation of the volume of a gas with the pressure and with temperature. To obtain the simultaneous variation of the volume with temperature and pressure, we proceed as follows. Consider a quantity of gas at P_1V_1 and T_1, and suppose that it is desired to obtain the volume of the gas, V_2 at P_2 and T_2. First of all, let us compress (or expand) the gas from P_1 to P_2 at constant temperature T_1. The resulting volume V_x then will be, according to Boyle's law,

$$\frac{V_x}{V_1} = \frac{P_1}{P_2}$$

$$V_x = \frac{V_1P_1}{P_2} \tag{18.6}$$

If the gas at V_x, P_2, and T_1 is heated (or cooled) now at constant pressure P_2 from T_1 to T_2, the final state at P_2 and T_2 will have the volume V_2 given by Charles' law, namely

$$\frac{V_2}{V_x} = \frac{T_2}{T_1}$$

$$V_2 = \frac{V_xT_2}{T_1}$$

Substituting into this relation the value of V_x from Eq. (18.6), V_2 becomes

$$V_2 = \frac{V_xT_2}{T_1} = \frac{P_1V_1T_2}{P_2T_1}$$

and on rearranging terms we see that

$$\frac{P_1V_1}{T_1} = \frac{P_2V_2}{T_2} = \text{constant} = K \tag{18.7}$$

In other words, the ratio PV/T for any given state of a gas is a constant. Consequently, we may drop the subscripts and write for any gas which obeys Boyle's and Charles' laws

$$PV = KT \tag{18.8}$$

Equation (18.8) is known as the combined gas law because it represents a combination of Boyle's and Charles' laws. It gives the complete relationship between the pressure, volume, and temperature of any gas as soon as the constant K is evaluated. That Boyle's and Charles' laws are merely special cases of Eq. (18.8) is easily shown. When T is constant, Eq. (18.8) reduces to PV = constant, or Boyle's law. Again, when P is constant, Eq. (18.8) becomes $V = (K/P)T = KT$, or Charles' law.

18.7 *The Gas Constant*

The numerical value of the constant K in Eq. (18.8) is determined by the number of moles of gas involved and the units in which P and V are expressed, but it is totally independent of the nature of the gas.* Equation (18.8) shows that for any given pressure and temperature an increase in the quantity of gas increases the volume and thereby also correspondingly the magnitude of K. In other words, K is directly proportional to the number of moles of gas involved. For convenience, this constant may therefore be replaced by the expression $K = nR$, where n is the number of moles of gas occupying volume V at P and T, and R is the gas constant per mole. Thus expressed, R becomes a universal constant for all gases, and Eq. (18.8) takes the final form

$$PV = nRT \tag{18.9}$$

Equation (18.9) is the ideal-gas equation, one of the most important relations in physical chemistry. It connects directly the volume, temperature, pressure, and number of moles of a gas, and permits all types of gas calculations as soon as the constant R is known. R may be found from the fact that one mole of any ideal gas at standard conditions (i.e., at 0°C and 1 atmosphere pressure) occupies a volume of 22.414 liters. If we express, then, the volume in liters and the pressure in atmospheres, R follows from Eq. (18.9) as shown below:

$$R = \frac{PV}{nT} = \frac{1 \times 22.414}{1 \times 273.16} = 0.08205 \text{ liter-atm/(degree)(mole)}$$

It should be clearly understood that although R may be expressed in various units, for pressure-volume calculations involving gases R must always be taken in the same units as those used for pressure and volume.

18.8 *Avogadro's Principle*

In 1811, Avogadro put forth the principle that equal volumes of all gases at the same pressure and temperature contain equal numbers of molecules. The actual number of molecules in a gram-mole of any gas is an important physical constant known as Avogadro's number, symbol N. This constant may be arrived at by a number of methods. The best present value for this quantity is 6.0235×10^{23} molecules per gram-mole. Once this constant is available, the mass of any particular

* A mole is the molecular weight of a material expressed in grams; that is, an amount weighing, in grams, the numerical amount of its molecular weight. Note—*molecular,* not atomic weight.

TABLE 18.1 Values of *U* in Various Units

Units of pressure	Units of volume	Temperature	n	R
Atmospheres............	liters	°K	g-moles	0.08205 liter-atm/(°K) (g-mole)
Atmospheres............	cu cm	°K	g-moles	82.05 cu cm-atm/(°K) (g-mole)
Dynes/cu cm............	cu cm	°K	g-moles	8.314×10^7 ergs/(°K) (g-mole)
mm Hg, (torr)...........	cu cm	°K	g-moles	62,360 cu cm-mm Hg/(°K) (g-mole)
Atmospheres............	cu ft	°R	lb-moles	0.730 cu ft-atm/(°R) (lb-mole)
Lb/sq in................	cu ft	°R	lb-moles	10.73 cu ft-(lb/sq in.)/°R (lb-mole)
Lb/sq in................	cu in.	°R	lb-moles	18.540 cu in.-(lb/sq in.)/ (°R) (lb-mole)
R, joules...............	....	°K	g-moles	8.314 joules/(°K) (g-mole)
R, cal..................	....	°K	g-moles	1.987 cal/(°K) (g-mole)

molecule can readily be computed by merely dividing the molecular weight of the substance by Avogadro's number.

18.9 *The Velocity of Gas Molecules*

According to the kinetic theory, all molecules at the same temperature must have the same average kinetic energy. It follows, therefore, that the higher the mass of a molecule the more slowly it must be moving. It is of considerable interest to ascertain the actual velocity with which various molecules move. From gas theory we have that

$$V = \left(\frac{3RT}{M}\right)^{1/2} \qquad (18.10)$$

where V = velocity, cm/sec
R = gas constant
T = absolute temperature
M = molecular weight

By this equation, the root-mean-square* velocity of a gas may be calculated from directly measurable quantities. In doing this, R must

* Root-mean-square velocity is the square root of the average of the squares of all the velocities involved.

be expressed in ergs per degree per mole, P in dynes per square centimeter, and the density in grams per cubic centimeter. With these units, V will be given in centimeters per second.

To calculate the velocity of hydrogen molecules at 0°C, we know that $R = 8.314 \times 10^7$ ergs per mole per degree, $T = 273.16$, and $M = 2.016$. Hence, Eq. (18.10) yields for V

$$V = \left(\frac{3RT}{M}\right)^{1/2}$$

$$= \left(\frac{3 \times 8.314 \times 10^7 \times 273.16}{2.016}\right)^{1/2}$$

$$= 184{,}000 \text{ cm/sec} = 68 \text{ mi/min}$$

Since hydrogen is the lightest of all elements, this tremendously high velocity represents an upper limit for rates of molecular motion. For all other molecules, the speeds will be lower. Thus for sulfur dioxide, with $M = 64$, the velocity at 0°C would be

$$\frac{V_{SO_2}}{V_{H_2}} = \left(\frac{M_{H_2}}{M_{SO_2}}\right)^{1/2}$$

$$V_{SO_2} = 68\left(\frac{2}{64}\right)^{1/2} = 12 \text{ mi/min}$$

chapter 19

Flow of Gases

In studying the flow of gases through the various parts of a vacuum system, three different flow regimes must be considered: viscous flow, molecular flow, and the regime between these two, transitional flow.

19.1 *Viscous Flow*

Viscous flow is that with which we are most familiar, being represented by the normal flow of water or compressed air in pipes and hoses. In it the fluid flows rapidly, and at low velocities rather uniformly, from regions of high pressure to those of lower pressure. This occurs because the molecules are in constant and rapid motion, colliding with one another, and bouncing off from the impact in all directions. However, in the downstream direction, where the pressure is lower, somewhat fewer molecules are present. The colliding molecules therefore suffer fewer collisions when moving in this direction, and move farther between collisions. The net effect is a rapid flow of molecules in the downstream direction.

For viscous flow of air at 25°C (viscosity = 1.845×10^{-4} poise) the conductance of a circular pipe is given by

$$C = 2.84 \frac{a^4}{L} \bar{P} \tag{19.1}$$

where C = tube conductance, liters/sec
a = radius of the tube, cm
L = length, cm
$\bar{P} = \dfrac{P_1 + P_2}{2}$, with P_1 = starting pressure and P_2 = final pressure

In English units this becomes

$$C = \frac{540D^4}{L} \bar{P} \tag{19.2}$$

where C = conductance, cu ft/min
D = diameter, in.
L = length, ft
$\bar{P} = \dfrac{P_1 + P_2}{2}$, torr

The English formula is much more convenient to use, since parts are normally dimensioned in feet and inches in fabricating shops in the United States, although not in Europe. In addition, the mechanical pumps sold in the United States are usually rated in cubic feet per minute, not in liters per second. Unfortunately, the manufacturers of diffusion pumps rate these in liters per second, even though the same firms make both. In consequence we must convert to these units when figuring molecular flow. It is unfortunate that the metric system cannot be used throughout, but at present this is not practical in shops operating in the United States.

In the above, we have used P_1 and P_2 to represent pressures at the beginning and end of pumpdown. Correctly stated, they should represent the pressures at the upstream and downstream ends of the pipe whose conductance is being measured. However, both change with the progress of the pumpdown and actually should be used in a series of computations for various segments of the pumpdown. In practice, however, the drop over the length of the connecting piping in a normal system will be small, and the two P's will move substantially together. The segments of the cycle are averaged by the technique used, with accuracy satisfactory for the great majority of cases. Where a long, low-conductance tube must be used, however, as in some evacuation methods, more detailed calculations are desirable.

19.2 *Mean Free Path*

As the pressure in a vacuum system is further and further reduced, the gas molecules tend to continue evenly distributed throughout the entire volume. Since the molecules are in rapid and constant motion, no concentration in any part of the system is even remotely probable. However, since the number of molecules is being reduced, the space between the molecules must increase correspondingly. This has the important result that the molecules' length of travel between collisions will also increase. The distance between collisions is known as the free path, or for a large number of molecules where the free paths vary statistically, as the "mean free path."

As the mean free path becomes longer and longer, a time comes when most or a great majority of collisions are with the walls of the confining vessel and no longer with other molecules. When this occurs, the molecules no longer show any tendency to flow in an orderly manner toward the pump but bounce aimlessly about within the vessel. Only those which happen by chance into the output opening are captured by the pump and removed. This is a condition known as "molecular flow."

As this condition is approached, a transition regime occurs, in which some viscous flow still occurs but so few molecule-to-molecule collisions occur that flow begins to approach that characteristic of molecular flow. Methods of computing conductance in the transition-flow regime are cumbersome and of low accuracy, since conditions change very rapidly. Fortunately, pressures pass through the transition range so rapidly under the usual conditions that the conductance in the transition range has a negligible effect on the pumpdown time. It will therefore be neglected herein.

The mean free path can be estimated for air at 70°F as follows:

$$L = \frac{1.91}{P}$$

where L = mean free path, in.
P = pressure, microns (1×10^{-3} torr)

Molecular flow can be assumed to begin when L is equal to or greater than the pipe radius in inches.

19.3 *Molecular Flow*

When the mean free path exceeds the radius of the conducting tube, most collisions of molecules will be with the walls of the tube, and not with each other. Under these conditions, the flow of gas will be wholly independent of the pressure and will depend only upon the radius of the tube, its length, the kinds of molecules flowing, and the temperature of the gas.

The best expression for the conductance of a tube under these conditions is that due to Clausing, as follows:

$$C = 2.64KA\left(\frac{T}{M}\right)^{1/2} \tag{19.3}$$

where C = conductance, liters/sec
A = area of tube, sq cm
T = temperature, °K
M = g/mole of gas (molecular weight)
K = a dimensionless function of length (given in centimeters) and radius (given in centimeters)*

* See Table 19.1 for values of K.

TABLE 19.1 Values of Clausing's Factor K

Length/radius (L/a)	K	Length/radius (L/a)	K
0.0	1.0000	3.2	0.4062
0.1	0.9524	3.4	0.3931
0.2	0.9092	3.6	0.3809
0.3	0.8699	3.8	0.3695
0.4	0.8341	4.0	0.3589
0.5	0.8013	5.0	0.3146
0.6	0.7711	6.0	0.2807
0.7	0.7434	7.0	0.2537
0.8	0.7177	8.0	0.2316
0.9	0.6940	9.0	0.2131
1.0	0.6720	10.0	0.1973
1.1	0.6514	12.0	0.1719
1.2	0.6320	14.0	0.1523
1.3	0.6139	16.0	0.1367
1.4	0.5970	18.0	0.1240
1.5	0.5810	20.0	0.1135
1.6	0.5659	30.0	0.0797
1.7	0.5518	40.0	0.0613
1.8	0.5384	50.0	0.0499
1.9	0.5256	60.0	0.0420
2.0	0.5136	70.0	0.0363
2.2	0.4914	80.0	0.0319
2.4	0.4711	90.0	0.0285
2.6	0.4527	100.0	0.0258
2.8	0.4359	1,000.0	0.002658
3.0	0.4205	α	$8a/3L$

By permission from Saul Dushman, "Scientific Foundations of Vacuum Technique," John Wiley & Sons, Inc., New York, 1962.

For area in English units,

$$C = 23.5KA\left(\frac{T}{M}\right)^{1/2} \tag{19.4}$$

where C = conductance, liters/sec
A = area, sq in.
T = temperature, °K
M = molecular weight (approximately 29 for air under standard conditions)
K = a dimensionless function of length (given in inches) and radius (given in inches)*

19.4 *Flow Conductance and Impedance*

Pumping speed of various types of vacuum pumps is given by the manufacturers as a function of pressure and as measured at the mouth of the pump. This should be carefully distinguished from the throughput, which is the volume pumped per unit time, multiplied by the pressure. Q is therefore proportional to mass, while speed is merely the volumetric flow rate and is not mass or proportional to mass until the pressure term is added.

Consider the following figure, representing a chamber connected to a pump by means of a pipe.

Let S_p = pump speed, measured at the mouth of the pump
C = conductance of pipe
S_n = pumping speed at the chamber wall

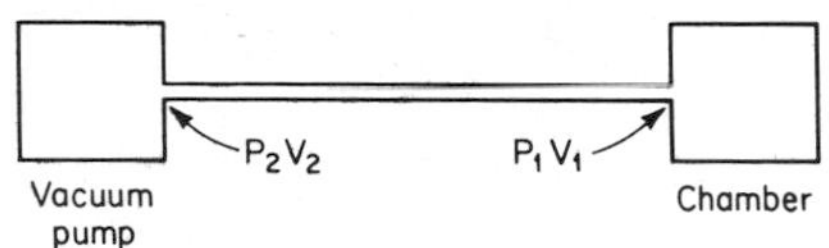

Since the pipe is continuous,

$$P_2V_2 = P_1V_1$$

Dividing by time t,

$$\frac{P_2V_2}{t} = \frac{P_1V_1}{t}$$

However, V/t is the pumping speed, by definition (volume per unit time). Therefore,

$$P_1S_n = P_2S_p$$

* See Table 19.1 for values of K.

However, the throughput Q is equal to PS and

$$Q = P_1S_n = P_2S_p$$

or

$$P_1 = \frac{Q}{S_n} \qquad P_2 = \frac{Q}{S_p}$$

Now conductance is the reciprocal of impedance, and

$$\frac{1}{C} = R$$

$$Q = \frac{P_1 - P_2}{1/C}$$

$$= (P_1 - P_2)C$$

Substituting,

$$Q = \left(\frac{Q}{S_n} - \frac{Q}{S_p}\right) C$$

Dividing by Q gives

$$1 = \frac{C}{S_n} - \frac{C}{S_p}$$

or

$$\frac{1}{C} = \frac{1}{S_n} - \frac{1}{S_p}$$

$$S_n = \frac{CS_p}{C + S_p} \qquad (19.5)$$

This is a most important relationship, which holds for both viscous and molecular flow. If, for example, we have a pump having a speed of 100 liters per second and connect it with a vessel to be evacuated by means of a connecting pipe having a conductance of 100 liters per second, the net pumping speed at the vessel will be

$$S_n = \frac{CS_p}{C + S_p}$$

$$= \frac{100 \times 100}{100 + 100} = \frac{10{,}000}{200} = 50 \text{ liters/sec}$$

In other words, the net useful pumping speed of the pump has been reduced 50 percent by using a connecting line having a conductance numerically equal to the speed of the pump. It is therefore *extremely* important to increase the conductance of the connecting lines between the pump and the vessel to the maximum extent practicable in order to achieve the greatest net speed obtainable. In practice, it is very difficult to achieve conductances equal to the speed of the pump at its mouth because of the molecular-flow regime; therefore, net speeds of 40 to 45 percent of the pump rating must usually be accepted.

In the viscous-flow regime, the reductions are usually negligible because of the inherent high conductivity of reasonable-sized pipes in this regime.

19.5 *Pumpdown Time in the Viscous-flow Range*

If a system could be constructed in such a way that no leaks were present and if no outgassing from walls, seals, and parts within the chamber were taking place, a simple formula would enable computation of the pumpdown time. The pumpdown time can be shown to be:

$$t = 2.3 \frac{V}{S} \log \frac{P_1}{P_2} \tag{19.6}$$

where t = time
V = volume of system
S = net pumping speed available
P_1 = starting pressure
P_2 = final pressure

This formula applies quite well during the roughing cycle, down to the pressure where outgassing from the walls begins, normally about 70 microns (7×10^{-2} torr). Outgassing becomes predominant below 10 microns (1×10^{-2} torr), and the formula no longer applies.

In using the formula to calculate the roughing time in a system, care must be taken to use compatible units. Where the net pumping speed of the mechanical pump and connecting piping has been given in cubic feet per minute, the volume must be in cubic feet, resulting in time in minutes. Where speed is in the units of liters per second, volume should be in liters, and time will be in seconds.

19.6 *Pumpdown Time in the Molecular-flow Range*

In the molecular-flow range, the net pumping speed usually used will be in the hundreds or thousands of liters per second. This gives rise to such large net speeds that even large chambers would be evacuated to very low levels in a time of a few seconds or less. This of course does not happen, for the reason that outgassing from the walls and work continually liberates more gas to replace that pumped out. The volume of gas so liberated may be thousands of times greater than the geometrical volume of the chamber. The time to reach a desired low pressure will therefore be the time when the net pumping speed becomes equal to the total of the leakage and outgassing at the desired pressure; that is,

$$P = \frac{Q_0 + Q_L}{S_n} \tag{19.7}$$

where Q_0 = outgassing rate divided by P
Q_L = leakage rate divided by P
S_n = net pumping speed at chamber wall at pressure P

In a well-constructed system, Q_L may be reduced to less than 1×10^{-9} atmospheric cubic centimeters per second for the entire system. This is so small that leakage seldom becomes the determinant in pumpdown time in a well-constructed system.

Q_0 may be estimated from the data on metals in Sec. 3.7 and on plastics from Sec. 3.8. The values given in these sections are not the initial outgassing rates for the materials involved, since initial outgassing rates occur at relatively high pressures in the viscous range and data are usually contaminated by chamber outgassing effects. Values are therefore given after 1 hour of pumping when rates are stabilized, at a constant slope versus time. It should be remembered that these rates decay with time—linearly for metals and at approximately the square root of time for plastics. This gives rise to the following table:

Pumping time, hr	*Stainless steel, % of base rate*	*Plastics, % of base rate*
1	100	100
2	50	70.7
3	33	57
4	25	50

It is obvious why plastics should be avoided in vacuum systems as far as possible. To determine the expected pumpdown time, it is convenient to subtract from the calculated net pumping speed at the chamber wall an amount equal to the leakage rate (determined by test using a calibrated leak and helium-leak detector). Thus

$$\overset{*}{S} = S_n - \frac{Q_L}{P_F} \tag{19.8}$$

where S_n = net pumping speed at chamber wall
Q_L = leakage rate, torr-liters/sec
P_F = final pressure desired

Then

$$t = \frac{Q_0}{\overset{*}{S}_n} \tag{19.9}$$

Time t will be in hours if Q_0 is in liters per second after 1 hour of pumping and $\overset{*}{S}$ is in liters per second at the final pressure.

When using any of the formulas for calculating conductance, pumpdown time, etc., it is important that consistent units be used throughout. Thus the constants of conversion may be useful: 28.3 liters = 1 cubic foot, and 1 liter per second = 2.12 cubic feet per minute. Mixing units can obviously lead to trouble! See Appendix A for a convenient list of conversion constants.

19.7 *The Importance of Properly Computing Net Pumping Speed*

A frequent error in the computation of pumpdown time arises from the assumption that the pumping speed available at the chamber is equal to the speed stated by the manufacturer to be available at his pump inlet. In the case of mechanical pumps, the pump speed decreases quite appreciably as the pressure goes down, so that one usually finds that at the lower pressures, where pumping is most difficult due to the beginning of outgassing effects, the net speed of the mechanical pump is at a minimum. It is therefore important to make use of the speed near the crossover pressure (where the diffusion pumping begins), as given by the pump manufacturer, in computing the speed of the pump at its inlet. Doing this instead of using the average speed provides a somewhat higher speed during the early portions of the pumpdown cycle than would be evident from the formula; however, the inclusion of outgassing effects due to water vapor at approximately 100 microns (1×10^{-1} torr) and below normally compensates for this slight error.

The impedance effect of the connecting piping between the mechanical pump and the chamber is not usually of great importance unless the lines are extremely long. In all cases, however, the conductance of the connecting piping should be checked by means of Eq. (19.2) to make sure that adequate pump speed at the chamber wall will be available.

For diffusion pumps, the situation is very different. Here we are dealing with molecular flow, and the impedance of the cold trap and connecting piping invariably lowers the available net pumping speed by a large amount. Referring to Eq. (19.5), it will be observed that if the conductance of the cold trap and connecting piping is equal to the speed of the pump in liters per second, the net pumping speed will only be one-half that of the pump alone. In practice, it is extremely difficult to design traps and connecting piping to provide a conductance equal to the speed of the pump at its connecting flange. In general, therefore, the net pumping speed at the chamber wall will range from 35 to 45 percent of the speed at the pump inlet.

19.8 *The Effect of Mixtures of Gases*

It will be noted from Eq. (19.3) that the molecular weights of the gases enter into the conductance formula as $(1/M)^{1/2}$. In the molecular-flow range, gas molecules rarely collide with each other. Each molecular species behaves, therefore, so far as flow is concerned, as if it were all alone in the system, being entirely uninfluenced by other gases present. Furthermore, ion pumps and diffusion pumps, as well as molecular-drag pumps, have different pumping speeds for different molecular species. Thus, in the case of diffusion-pumped systems, both the speed of the pump and the conductance of the connecting piping are much higher for hydrogen than for oxygen or water vapor, and the diffusion pump will therefore handle a larger volume of this gas. The backing mechanical pump, however, being a volumetric-displacement device, makes no distinction between gases and pumps the same number of cubic feet per minute of hydrogen as of anything else.

In extreme cases where very large amounts of hydrogen are evolved (such as from large amounts of highly heated graphite within the chamber), the backing pump may be unable to handle the hydrogen pumped by the diffusion pump at a sufficiently rapid rate. The result can be a rapid rise of the diffusion pump forepressure to a value causing vapor-stream breakdown in the upper jet, resulting in massive oil backstreaming into the hot chamber. In cases where such hydrogen evolution is to be expected, the effect should be calculated and extra-large backing capacity provided.

A more common effect of gas composition is found in ion-gauge tubulation. If the ion gauge has an appreciable pumping speed of its own, which is usually the case, gas will flow from the system into the gauge tube. Since the conductance of the tubulation is different for each type of molecule, and since the actual pumping speed of the gauge is different for each gas, the effect will be to create a gas composition within the gauge different from that within the chamber. Also, the total pressure within the gauge will be different from and lower than that within the chamber, due to the impedance drop through the tubulation. Also, the gauge response differs for different gases, as is explained in Chapter 9.

It follows from the above that a diffusion pump attached to a system having a net pumping speed for air of 2,000 liters per second may pump 2,000 liters per second of air, while at the same time pumping an additional 2,000 liters per second of hydrogen. Meanwhile a nude (untubulated) ion gauge attached to the system will actually read lower after the addition of the hydrogen leak than before. The effect of the pump should be understandable from the manner in which gases flow in the molecular-flow regime. The gauge effect is described more fully in Chapter 9.

chapter 20

Pumping Calculations

20.1 *Introduction*

In Chapters 18 and 19, methods of computing the flow of gases through connecting piping, computing net pumping speed, etc., were outlined in some detail. Having also described methods of fabricating vacuum systems, types of pumps, and accessories employed, we will attempt to illustrate, by example, some of the calculations involved in practical vacuum-system design.

20.2 *Selection of Pumps for a Small Bell-jar-type Vacuum Coater*

GIVEN: An 18-inch-diameter by 24-inch-high stainless steel bell jar and base to be used for vacuum coating. Room temperature: 27°C. Backfilling gas: Room air (molecular weight = 29).

REQUIRED: Pumpdown time to 1×10^{-5} torr: 60 minutes.

DETERMINE: Size of mechanical pump required; size of diffusion pump required; size of connecting piping, valves, and traps.

SOLUTION: Make simplifying assumptions:

1. Roughing time approximately 15 minutes.
2. Diffusion-pumping time approximately 45 minutes.
3. Diffusion pump will be heated and held on holding pump until roughing is complete, so no allowance is necessary for warmup.
4. Crossover pressure is 10 microns (1×10^{-2} torr) to minimize oil backstreaming.

1. *Mechanical Pump*

$$S_N = \frac{2.3V}{t} \log \frac{P_1}{P_2} \qquad (20.1)$$

where $V = (1.5)^2 \times \frac{\pi}{4} \times 2 = 3.52$ cu ft

P_1 = atmosphere, taken as 1,000 torr

$P_2 = 1 \times 10^{-2}$ torr

$$S_N = \frac{2.3 \times 3.52}{15} \times \log \frac{10^3}{10^{-2}}$$

$$= 0.54 \times 5 = 2.7 \text{ cu ft/min}$$

Pick a pump having a speed at 10 microns (1×10^{-2} torr) of 2.7 cubic feet per minute or more. A suitable pump is a Kinney KC5 or KCV5 (two-stage), giving in either case a speed at 10 microns (1×10^{-2} torr) of 3.5 cubic feet per minute.

Using this pump, roughing time

$$t = \frac{2.3 \times 3.52}{3.5} \times \log \frac{10^3}{10^{-2}} = 11.6 \text{ min}$$

2. *Low-pressure Piping* Pump inlet = 1 inch N.P.T. Assume a length of roughing line, including valve, of 3 feet and a diameter of 1 inch.

$$C = \frac{540D^4}{L} \times \bar{P} \qquad (20.2)$$

$$C = \frac{540 \times 1^4}{3} \times \frac{760 + 10^{-2}}{2} = \frac{540}{3} \times 380 = 68{,}000 \text{ cu ft/min}$$

This conductance is so large over the average pressure range during pumpdown as to present no appreciable impedance. The mean free path

$$L = \frac{1.91}{P_{\text{micron}}} = \frac{1.91}{10} = 0.191 \text{ in.}$$

Therefore, the flow is viscous during the entire roughing cycle, and Eq. (20.2) applies.

3. *Diffusion Pump and Connecting Piping* At the ultimate pressure required in the system, the net diffusion-pumping speed must equal the rate of gas evolution from the walls divided by the pressure, or

$$S_N = \frac{Q}{P}$$

where Q is the wall area multiplied by the outgassing rate.

$$\begin{aligned} \text{Area} &= 1.5 \times \pi \times 2 = && 9.40 \text{ sq ft} \\ &+ (1.5)^2 \times \frac{\pi}{4} \times 2 = && 3.54 \text{ sq ft} \\ \text{Total} & && \overline{12.94} \text{ sq ft} \end{aligned}$$

The outgassing rate for stainless steel is 0.16×10^{-3} torr-liters per second after 1 hour of pumping for an unbaked system.

$$S_N = \frac{12.94 \times 0.16 \times 10^{-3}}{1 \times 10^{-5}} = 206 \text{ liters/sec}$$

after 1 hour of pumping.

Assume a P.M.C. 720 (nominal 4-inch) diffusion pump having a speed of 720 liters per second at the pump flange.

Connecting tubulation (for 4-inch diameter line):

Base plate and tube to flange, in.	3.0
Gate valve (flange-to-flange), in.	4.5
Chevron baffle 5.5-inch flange-to-flange plus path length increase due to chevrons, in.	6.0
Total length, in.	13.5

Conductance for molecular flow will be

$$C = 3.64 \times KA \left(\frac{T}{M}\right)^{1/2} \tag{20.3}$$

Now we have

$$\begin{aligned} L &= 13.5 \text{ in.} = 34.2 \text{ cm} \\ a &= 2 \text{ in.} \times 2.54 = 5.1 \text{ cm (radius of pipe)} \\ \frac{L}{a} &= \frac{34.2}{5.1} = 6.6 \end{aligned}$$

From Table 5.1

$$\begin{aligned} K &= 0.27 \\ T &= 273°\text{C} + 27°\text{C} = 300°\text{K} \\ M &= 29 \end{aligned}$$

Therefore

$$A = \pi \times (5.1)^2 = 80 \text{ sq cm}$$

$$C = 3.64 \times 0.27 \times 80 \times \left(\frac{300}{29}\right)^{1/2}$$

$$= 3.64 \times 0.27 \times 80 \times 3.3 = 260 \text{ liters/sec}$$

$$S_N = \frac{S_P C_L}{S_P + C_L} = \frac{720 \times 260}{720 + 260} = 190 \text{ liters/sec}$$

Thus the time required for pumpdown will be

$$t = 1 \text{ hr} \times \frac{206}{190} = 1.08 \text{ hr} = 65 \text{ min}$$

This is too long to meet requirements. Let us try again, using a 6-inch-diameter connecting line.

$$L = 34.2 \text{ cm}$$
$$a = 3 \times 2.54 = 7.6 \text{ cm}$$
$$\frac{L}{a} = \frac{34.2}{7.6} = 4.5$$
$$K = 33$$
$$A = (7.6)^2 \times \pi = 180 \text{ sq cm}$$

Therefore

$$C = 3.64 \times 0.33 \times 180 \times \left(\frac{300}{29}\right)^{1/2} = 720 \text{ liters/sec}$$

$$S_N = \frac{720 \times 720}{720 + 720} = 360 \text{ liters/sec}$$

$$t = 1 \text{ hr} \times \frac{206}{360} = 0.57 \text{ hr} = 34.2 \text{ min}$$

Total pumpdown time:

Roughing, min.	11.6
Diffusion pumping, min. . . .	34.2
Total, min.	45.8

This illustrates the importance of having a large enough line between the diffusion pump and chamber so that speed is not unduly choked down.

However, it should be noted that the outgassing rate used was given after 1 hour of pumping, while in the case used the total pumping time indicated is only 45.8 minutes. Since the rate decays with time, it will be larger at the end of 45 minutes, and the actual pumpdown time would be expected to be substantially longer. We would therefore estimate that the first pumpdown will take approximately 1 hour. Subsequent pumpdowns might closely approach the theoretical 45 minutes, provided backfilling is done with dry air or dry inert gas and provided that the open period is kept small.

In actual operation, numerous auxiliary items will be present in the chamber, including evaporative filaments or boats, substrate material, masks, and manipulative devices. All of these will contribute to the outgassing effects, as will the small amount of leakage associated with the various mechanical and electrical passthroughs. On the other hand, the wall outgassing will gradually decay from the values used above with additional pumping time, and will not return to this value provided backfilling is accomplished with dry air or dry inert gas. The net result will be that the normal pumpdown time for successive coating runs can be kept near the 45-minute time used in the calculations until the bell jar becomes so seriously contaminated with coating materials, organics from the pump fluid, etc., as to require a major cleaning operation.

4. *Backing-pump Requirements* So far, we have tacitly assumed that a mechanical pump of sufficient size for roughing purposes would be of sufficient size to back the diffusion pump. It now becomes necessary to check this assumption.

We note from the characteristic curves of the Bendix Corporation P.M.C. 720 diffusion pump that the throughput at 1×10^{-2} torr is 2,200 micron-liters per second (220 liters per second speed multiplied by 1×10^{-2} torr). To avoid backstreaming of a massive sort due to breakdown of the jets, the forepressure of the pump must be maintained at 400 to 450 microns (4 to 4.5×10^{-1} torr). The mechanical pump must therefore handle 2,200 micron-liters per second at a pressure of, say, 400 microns (4×10^{-1} torr). The required pumping speed will be

$$S_P = \frac{2200 \times 10^{-3} \text{ liters/sec}}{400 \times 10^{-3}}$$
$$= 5.5 \text{ liters/sec}$$
$$= 11.7 \text{ cu ft/min}$$

Referring to the Kinney pump curves again, we see that this will require the use of a KC-15 pump, which is a larger size than the size previously selected. This can only be avoided by accepting a certain amount of backstreaming during the cutover period, as is usually done, or by using an ejector stage between the mechanical pump and the diffusion pump. This latter method is normally used on larger systems, but is not economical on a system as small as this example.

The mechanical holding pump, required to hold the blanked-off diffusion pump during roughing of the chamber, may be of the smallest size available, since the blanked-off diffusion pump will have an extremely small throughput, even during startup. In this case, the holding pump might well be a Kinney CV-2.

20.3 *Calculating a System Containing Both Plastic and Metal Parts*

GIVEN: A 24-inch-diameter by 36-inch-long horizontally opening chamber fabricated of stainless steel.

LOAD: A cylindrical tube of stainless steel, 6 inches in diameter by 24 inches long, to which are affixed 4 plastic disks (Plexiglas), each 24 inches in outside diameter (O.D.) by ½ inch thick, with central holes permitting the disks to be spaced apart on the 6-inch mandrel.

The chamber has been preconditioned by baking and has an outgassing rate, after conditioning, of 1×10^{-5} torr-liters per second, decaying linearly with time.

The outgassing rate of the mandrel is 0.16×10^{-3} torr-liters per second per square foot, decaying linearly with time.

The outgassing rate of Plexiglas is 2×10^{-3} torr-liters per second, decaying as the square root of the time ($t^{1/2}$).

DETERMINE: Pumpdown time to 1×10^{-5} torr.

SOLUTION: From a quick inspection of the outgassing rates, it is apparent that the pumpdown time will be almost completely determined by the diffusion-pump net speed. We should therefore use the largest diffusion pump which can be conveniently attached to the chamber. A comparison of the heights of various diffusion pumps and of the probable lengths of elbows and traps indicates that the use of any size larger than a 10-inch pump will require the provision of a pit or elevation of the chamber to an inconvenient height. We therefore choose a Varian HS 10-4200 pump having an unbaffled speed of 4,200 liters per second and a speed with liquid nitrogen baffle of 2,000 liters per second at 1×10^{-5} torr. The manufacturers' literature states that a 20 cubic foot per minute backing pump is required for full throughput.

1. *Roughing Time (Neglecting Line Impedance)* A suitable backing pump would be the Kinney KC-46, having a speed at 10 microns

(1×10^{-2} torr) of 32 cubic feet per minute. Roughing time would thus be

$$t = 2.3 \frac{V}{S} \log \frac{P_1}{P_2}$$

$$V = (2)^2 \times \frac{\pi}{4} \times 3 = 9.4 \text{ cu ft}$$

$$t = \frac{2.3 \times 9.4}{32} \times \log \frac{1{,}000}{1 \times 10^2}$$

$$= \frac{2.3 \times 9.4 \times 5}{32} = 3.4 \text{ min}$$

2. *Time to Pump from* 1×10^{-2} *Torr to* 1×10^{-6} *Torr* Outgassing:

Walls:

$$\frac{A \times 1 \times 10^{-5} \text{ torr-liters/sec}}{t}$$

$$A = \pi \times 2 \times 3 = 18.7 \text{ sq ft}$$

$$+ \frac{\pi}{4} \times 2^2 \times 2 = 6.3 \text{ sq ft}$$

$$\text{Total } A = 25.0 \text{ sq ft}$$

$$Q_w = \frac{25 \times 1 \times 10^{-5}}{t} = \frac{2.5 \times 10^{-4}}{t}$$

Mandrel:

$$\pi \times \frac{6}{12} \times 2 = 3.34 \text{ sq ft}$$

$$Q_m = \frac{0.16 \times 10^{-3} \text{ torr-liters/(sec)(sq ft) } 3.34 \text{ sq ft}}{t}$$

$$= \frac{5 \times 10^{-2}}{t} \text{ torr-liters/sec}$$

Plexiglass:

$$Q_p = A_p \times \frac{2 \times 10^{-3}}{t^{1/2}}$$

$$A = (2)^2 \times \frac{\pi}{4} \times 2 \text{ sides} \times 4 \text{ disks}$$

$$= 25 \text{ sq ft}$$

$$Q_p = \frac{25 \times 2 \times 10^{-3}}{t^{1/2}} = \frac{5 \times 10^{-2}}{t^{1/2}}$$

Total:

$$Q_t = Q_w + Q_m + Q_p$$

$$= \frac{2.5 \times 10^{-4}}{t} + \frac{5 \times 10^{-2}}{t} + \frac{5 \times 10^{-2}}{t^{1/2}}$$

Pump speed at upper flange of cold trap is 2,000 liters per second. Use a connecting tube, say, 12 inches in diameter by 4 inches long, which is to say 30.5 cm in diameter by 10 cm long.

$$\frac{L}{a} = \frac{10}{15.3} = 0.65$$

$$K = 0.755 \text{ from Table 5.1}$$

$$C = 3.64\,KA \times \left(\frac{T}{M}\right)^{1/2}$$

We have

$$T = 300°\text{K}$$
$$M = 29$$
$$A = \pi \times (15.3)^2 = 7.35 \text{ sq cm}$$

Therefore,

$$C = 3.64 \times 0.755 \times 735 \times \left(\frac{300}{29}\right)^{1/2}$$
$$= 3.64 \times 0.755 \times 735 \times 3.3 = 6{,}500 \text{ liters/sec}$$
$$S_N = \frac{6{,}500 \times 2{,}000}{6{,}500 + 2{,}000} = 1{,}530 \text{ liters/sec}$$

We now equate the pumping speed to the total outgassing and solve for t:

$$1{,}530 \times 1 \times 10^{-5} = \frac{2.5 \times 10^{-4}}{t} + \frac{5 \times 10^{-2}}{t} = \frac{5 \times 10^{-2}}{t^{1/2}}$$

Reducing exponents to 10^{-3},

$$15.3 \times 10^{-3} = \frac{0.25 \times 10^{-3}}{t} + \frac{50 \times 10^{-3}}{t} + \frac{50 \times 10^{-3}}{t^{1/2}}$$

Multiplying through by 10^3t, we get

$$15.3t = 0.25 + 50 + 50(t)^{1/2}$$

Rearranging,

$$15.3t - 50(t)^{1/2} - 50.25 = 0$$

Use quadratic equation

$$x = \frac{-b \pm \sqrt{b^2 - 4ac}}{2a}$$

and solve for positive root only:

$$x = (t)^{1/2} = \frac{-(-50) \pm \sqrt{(-50)^2 - 4 \times 15.3 \times (-50.25)}}{2 \times 15.3}$$

$$= 50 \pm \sqrt{\frac{2{,}500 + 3{,}100}{30.6}}$$

$$= \frac{50 + 75}{30.6} = 4.1 \text{ hr}$$

$$t = x^2 = (4.1)^2 = 16.7 \text{ hr}$$

It will be seen that the roughing time of 3.4 minutes is negligible and that the diffusion-pumping time can be considered as representing practically the total pumping time in systems where heavy outgassing is encountered.

20.4 *Designing an Ultrahigh Vacuum System*

In designing a chamber capable of operation in the ultrahigh-vacuum region (i.e., at pressures of 1×10^{-10} torr or below), the same basic methods of computation exemplified above would be used. However, at these low pressures, certain changes in system construction are required.

1. Bakeout of the system will be required, at temperatures of 300°C or higher. This requirement dictates that the system be fabricated of stainless steel throughout and that all gaskets, seals, and valves be of the all-metal type to withstand bakeout temperatures and repeated heating and cooling. A heater system with suitable insulation, pyrometric control system, and cooling means should be a basic part of the design.

2. Leak checking of the system must be meticulous, with *all* detectable leaks removed. The minor, undetectable leaks should be measured by enclosing the entire system in a plastic sheeting bag, filled with helium. Total leakage should be below 1×10^{-9} standard cubic centimeters per second in order to avoid undue loading of the pump by this source.

3. Gauges capable of accurate reading of the pressures involved must be provided. In general, this precludes the use of Bayard-Alpert hot-filament gauges. Tubulated gauges of this type tend to be inaccurate at pressures below 1×10^{-9} torr, and the nude variety at pressures below 3 to 5×10^{-10} torr. The most suitable gauges for this pressure range are the cold-cathode magnetron or Penning types, especially those with a metal case and high-conductivity tubulation, which may be most read-

ily baked out. The General Electric triggered Penning gauge and the Varion extractor gauge are the most satisfactory currently available.

4. Backstreaming of diffusion pump fluids must be completely eliminated through the use of very stable, low-vapor-pressure pump fluids and high-conductance, "double-bounce" liquid-nitrogen-cooled traps.

5. The total available pumping speed must be very high, since outgassing effects at these pressures represent a very large volumetric load.

Consider the 24-inch-diameter by 36-inch-long chamber described above in Sec. 20.3, modified to eliminate the elastomeric seals on the door, passthroughs and cold trap, and the elimination of the elastomerically sealed gate valve between the trap and chamber. A second diffusion pump of 4-inch nominal diameter will be provided between the main diffusion pump and the mechanical pump. This second stage diffusion pump serves several purposes:

1. It allows bakeout of the chamber *and* cold trap with the main pump cut off, thus cleaning the trap while still preventing diffusion-pump oil from reaching the chamber.

2. It retards backdiffusion of hydrogen from the breakdown of mechanical-pump oil, which can occur when only a single diffusion stage is provided.

3. It maintains a backing pressure of 1×10^{-6} to 1×10^{-7} torr for the main diffusion pump, thereby aiding the latter in pumping to very low pressures.

4. It allows the main pump to be energized at a system pressure of 1×10^{-6} torr or lower, leading to more rapid stabilization of the jets and lessening backstreaming during the critical stage before jet action is fully established.

The outgassing rate from the chamber walls can be reduced to a value of approximately 1×10^{-5} torr-liters per square foot per second after a 24-hour preconditioning run followed by opening to air. A short bake of 4 to 6 hours, running on the backing diffusion pump, will normally reduce the wall outgassing to approximately 1×10^{-7} torr-liters per square foot per second. With a wall area of 25 square feet for the referenced chamber, this will represent a load of 25×10^{-7} or 25,000 liters per second required pumping speed at 1×10^{-10} torr.

The net pumping speed of the 10-inch diffusion pump can be held to approximately 1,800 liters per second by very careful attention to tubulation design and by using an oversized elbow-type trap. The pumpdown time will therefore be

$$t = \frac{25 \times 10^{-7} \text{ torr-liters/sec}}{1{,}800 \text{ liters/sec pump speed}} \times 1 \text{ hour} = 14 \text{ hours}$$

Total pumpdown time:

Diffusion pumping, hr.....	14.0
Baking, hr...............	6.0
Cooling, hr...............	2.0
Roughing, hr.............	0.5
Total, hr...............	22.5

The outgassing decay rate is now very slow and no longer linear with time. In general, equilibrium will be reached in about an additional 24 hours at a value of 3×10^{-11} torr or slightly lower.

The above calculation holds for a clean, dry, and empty system. Actual systems, however, must provide for handling outgassing from test articles or work placed within them. It is rather obvious that the addition of any appreciable gas load to that of the chamber walls will seriously degrade the vacuum capability, since no excess pump speed is available. A chamber of this size with this pumping capability will normally operate at 1×10^{-7} to 1×10^{-8} torr with actual loads.

A larger diffusion pump might seem the logical way to improve conditions. However, for a 24-inch-diameter chamber, the use of the next-largest-sized pump (18 inches nominal diameter, 20.4 inches I.D. of flange) is impractical unless the cold trap elbow is so reduced in size as to reduce the theoretical pump speed to a value little better than that of the 10-inch pump.

An ion-sublimation pump, which requires no cold trap or elbow, can be used; this pump will give a step-up in net pumping speed by a factor of approximately 2.5. This is a help, but not enough. Additions to the pump speed of decades are needed.

The most effective way of increasing pumping speed is to resort to cryopumping within the chamber, using cold (18°K) helium gas as the cooling medium. It was shown in Chapter 10 that the cryopumping speed of an array is

$$S_a = 10{,}800G\left(\frac{28}{M}\right)^{1/2}$$

For air, and with a G value of 0.25, this becomes $S_a = 10{,}800 \times 0.25 \times 1 = 2{,}700$ liters per second per square foot.

For a 24-inch-I.D. chamber, the shroud I.D. will be approximately 14 inches and the length 30 inches. The area is therefore

$$\pi \times \frac{14}{12} \times 2.5 = 9.2 \text{ sq ft}$$

$$S_a = 9.2 \times 2700 \text{ liters/sec} = 24{,}600 \text{ liters/sec}$$

This will handle outgassing from the load at 1×10^{-10} torr if $Q = 24{,}600 \times 1 \times 10^{-10}$ torr $= 2.5 \times 10^{-6}$ torr-liters/second.

This is sufficient to maintain a pressure of 1×10^{-10} torr for many types of loads run in such a chamber, making it useful for ultrahigh-vacuum work; whereas without such pumping capability, it remains only an interesting laboratory curiosity, or must be operated at some higher pressure.

chapter 21

Uses of Vacuum Systems

21.1 *Introduction*

Vacuum is used in so many different applications, both in production and in research, that a complete description of all the uses would entail a catalog longer than the entire size of this book. In the following, however, we will try to describe some of the most important fields in which vacuum is being used, either in a research sense or in a production sense, and show illustrations of some of the most common equipment employed in these uses. A large number of manufacturers produce equipment of these types, and the use of illustrations from a given company does not imply that only the company cited produces equipment of this type. There are in addition many specialized applications answering particular production or research needs, which have been developed to a state of useful perfection by those having need of such equipment. The fields covered herein are those having more general application and will be more frequently encountered in practice.

21.2 *Vacuum Metallizing Equipment*

Vacuum coaters, evaporators, or metallizers, to use three terms having the same connotation, are vacuum chambers in which items are coated with an evaporated film of metal or nonmetal to produce some particular effect. This field is by far the most rapidly growing of all the fields of vacuum application, and probably accounts for 40 percent of the total usage of vacuum pumps and equipment at the present time. System sizes range from very small bell jars capable of taking a single small specimen to systems large enough to handle the 200-inch mirror used on Mt. Palomar in the Hale telescope.

From the volume standpoint, the greatest amount of use of this process is in the coating of plastic, base metal, or ceramic objects with coatings of aluminum to produce high reflectivity and simulated noble-metal appearance. Thus plastic toys, decorative jewelry, decorative lamp fixtures, and many items of an ornamental nature are fabricated out of low-cost materials such as plastics or zinc die-castings, and then, after proper preparation, are coated with a thin film of aluminum. In the case of many items a coat of lacquer is placed on the article before coating takes place. After coating, such items are usually overcoated with a second coat of lacquer to protect the aluminum from corrosion and quick destruction. Such overcoats may be either of clear lacquer, when silver color is desired, or may be colored, when copper, gold, or some other color is desired. Very large numbers of such small items are suspended from movable racks, placed within vacuum chambers, and arranged in such a way that they may be rotated during evaporation to secure uniform and continuous coating of the article.

The evaporant consists of loops of aluminum foil, or wire, suspended from tungsten filaments, in such a fashion that the foil can be vaporized after the system has been pumped down to an appropriate pressure. In cases where a large amount of evaporant is required, a tungsten boat containing aluminum is provided, which is heated either by resistance heating or electron-beam heating at the appropriate period and for a suitable interval. The entire process is completed very rapidly, the principal element of time being the time to pump down the system.

Pressures required for aluminum-coating items of an ordinary nature are on the order of 1×10^{-3} or 1×10^{-4} torr, and there is no necessity for pumping to lower values for this type of work.

The coating of front-surface mirrors for optical purposes is carried out in the same fashion, except that in the case of mirrors it is generally necessary to provide a stronger overcoat, and one with less light absorption, than the lacquer used on ornamental items. In this case, an overcoat is formed by evaporating silicon monoxide from a resistance-heated boat, or one heated by an electron beam, which deposits the silicon monoxide on top of the aluminum coating in a thin and nearly impervious layer. A system pressure of approximately 1×10^{-5} torr is required. After exposure to the oxygen in the air, such coatings convert to a transparent layer of silicon dioxide, which provides greater abrasion resistance than the silicon monoxide coat but remains clear and transparent.

Coatings of metals other than aluminum are readily achievable, and such metals as gold, silver, and nickel are frequently deposited for special applications, gold being especially desirable for reflectors used in the infrared region of the spectrum, where silver and aluminum are inferior to it.

The total amount of metal deposited by these procedures is very small indeed, since the deposited film in the usual case has a thickness of only 10^{-6} inch, so that the cost of metal for the metallizing operation is very nominal. It should be recognized, however, that the vapor stream coming from the hot surfaces moves in straight lines in all possible directions, so that not only the desired articles are coated but all the interior of the chamber as well, although the only useful application is to the desired substrate. Shadow effects due to uneven coating, or oblique-angle effects, take place whenever the jet vapor stream impinges the surface at an angle differing greatly from 90 degrees. This property is made use of when "shadow casting" replicas of surfaces to be viewed in the electron microscope where the shadow effect outlines surface irregularities not otherwise visible.

Figure 21.1 shows typical small bell-jar (diffusion-pumped) coaters of the type used for small objects. Figure 21.2 shows a similar system

Fig. 21.1 Diffusion-pumped bell-jar coaters.

Fig. 21.2 Ion-pumped bell-jar coater, showing sorption roughing pumps.

utilizing ion pumping. Figure 21.3 shows a system with rotating rack, used for coating large numbers of toys or decorative elements. Figure 21.4 shows a 5-foot-diameter by 5-foot-high unit used for aluminizing 5-foot-diameter mirrors. Figure 21.5 shows the chamber open with mirror in place, and Fig. 21.6 shows a close-up of the filaments and boats used in this system.

In addition to evaporating the materials by resistance heating of the filaments or boats, electron-beam bombardment is frequently used, particularly when the material to be evaporated has an extremely high melting point or where it would react with the boat or filament material.

From a vacuum-system standpoint, the coating chambers so far considered are high-production devices, where the speed of pumpdown is a primary consideration. Consequently, they are often provided with very large diffusion pumps, designed so that they can be brought on line at a relatively high pressure, although they are not often equipped with liquid-nitrogen traps of an elaborate design. Instead, the emphasis is on high conductivity, to enable pumpdown in the shortest possible

Fig. 21.3 Rotating rack for decorative coater.

Fig. 21.4 Coater for 5-foot-diameter mirrors.

Fig. 21.5 Coater with mirror in place.

time. A shutoff valve and holding pump are provided, so that the diffusion pump may be left on line while the work is being removed and replaced.

Often a power supply for high voltage and suitable electrodes are provided, so that molecular bombardment may be used during pumpdown to remove gases and greases on the surface of the work, thus shortening the pumpdown period and cleaning the surfaces for the reception of the evaporate. Fully automatic systems are usually provided, so that operators need not have a high degree of skill nor long training to turn out acceptable work.

21.3 *Evaporators Used in Electronic-circuit Manufacture*

In the manufacture of microminiature circuitry, including integrated circuitry, transistors, diodes, etc., it is possible to build up various circuit elements and interconnections on small glass, quartz, or silica chips by evaporative techniques. A metal mask is used which permits deposition only on selected areas of the surface. A succession of such masks is used to deposit resistors, capacitors, connecting leads, and insulators, all on a chip so small that optical aid is required to observe the indi-

Fig. 21.6 Filaments and boats for coater.

vidual elements. By this means hundreds of elements of extremely small size can be deposited in one unbroken series of operations, so that circuits performing very elaborate functions can be reduced to the size of a dime or even smaller.

Vacuum systems for this type of work are not usually of large size but are very complicated due to the large number of penetrations required. In the first place, there must be one or more sources for each of the materials to be evaporated. This might include aluminum, graphite, nickel-chromium, silver, gold, silicon monoxide, and a number of other specialized materials. The arrangement may use alloy wires or, more usually, boats or crucibles heated by resistance heaters of tungsten or molybdenum or by electron beams as may be appropriate.

Provision must be made to store all the masks to be used inside the chamber and to position them over the substrate in the desired order, holding positional errors to 0.001 inch or less. Provision must be made to heat or cool the substrate—or both alternately—to a precise temperature. Cleaning means by molecular bombardment must be available if needed, and in addition, extremely uniform coating must be designed for, by means of rotation of the source or the substrate (including masks), or by other means. It is also necessary, in most systems, to have manipulative arms project within the chamber, provided with 3 degrees of freedom, in order to permit mechanical operations inside the chamber without breaking vacuum.

It is necessary, in addition to the above, to provide readout devices to indicate pressure and temperature of all critical devices and monitors which provide a readout to the operator of the rate of deposition and total thickness deposited. Usually a special passthrough collar is provided which gives additional room for electrical and mechanical penetrations.

The vacuum systems proper used in evaporators of this type differ from those described earlier chiefly in the requirements for great cleanliness. In these systems, therefore, highly sophisticated cold traps using liquid-nitrogen cooling are usually used, and every possible effort is made to avoid contamination of the substrates by organic materials. Pressure levels used range from 1×10^{-4} to 1×10^{-8} torr and occasionally lower in special cases. In some cases molecular sieve pumps are used for roughing, and ion pumps and titanium sublimator pumps are used to completely eliminate organic contamination.

Unfortunately, the thermal method of evaporation so far described has certain weaknesses. When it is desired to deposit an alloy, such as nickel-chromium for a resistor, composition of the deposit soon gets out of hand. The evaporation from an alloy tends to occur first from the element having the lowest melting point, with the result that the resulting deposit is enriched undesirably in this element. It is possible to codeposit from two sources simultaneously, each constituting one of the metals composing the alloy. However, composition control continues to be very difficult by this method.

21.4 *Sputtering Methods*

For the above and other reasons, a method known as sputtering is often used. In this method, the chamber is first evacuated in the ordinary way to a pressure of 1×10^{-6} torr, then backfilled, usually with purified argon, to a pressure of 10 to 12 microns (1 to 1.2×10^{-2} torr). In a part of the chamber hidden by a baffle, a beam of electrons from a hot filament is accelerated and used to ionize some of the argon. The argon ions are then extracted, moved into the chamber, and caused to impact negatively charged plates of the materials to be deposited. The accelerating voltage used is sufficient to cause the impinging ions to sputter material from the plates. By proper geometrical arrangement, the neutral sputtered molecules or particles are caused to deposit on the mask-covered substrate. The advantage of this method is that the sputtering rate of the various elements in an alloy is not proportional to their melting point, as in an alloy being thermally evaporated, but only weakly to their mass. By suitable mechanical arrangements, the plates to be sputtered can be changed, and various materials deposited.

This technique has some unique advantages for microcircuitry applications, especially where alloys must be deposited, a familiar example being the nickel-chromium resistor material previously mentioned. Unfortunately, contamination from any impurities present in the chamber or in the gas supply is quite probable because of the relatively high pressure.

Where sputtering is used for deposition, pressures of 10 to 20 microns (1 to 2×10^{-2} torr) are common in order to permit creation of the large number of ions required. However, removal of the active gases, nitrogen, oxygen, hydrocarbons, and water must be quite complete prior to introduction of the working gas, usually argon. Where oxidation is desired during deposition, purified oxygen is admitted in controlled amounts along with the argon. In all cases, therefore, a good high-vacuum system is required to evacuate the system to remove the contaminants before the admission of the working gas.

One of the problems inherent in all coaters is that the material being evaporated or sputtered not only coats the desired substrate but all the parts of the chamber as well. This includes the windows through which the operator may wish to observe the operation. The most common way of avoiding this problem is to provide a small shutter inside the system which, when closed, will shield the window from the evaporant. The shutter may be operated by a shaft extending through the chamber wall or by an internal solenoid controlled electrically from outside. Unfortunately, the shutter must remain closed during the actual evaporating cycle, which is precisely the time when vision is most desired.

Another method is to use two spools with a roll of Mylar tape arranged so as to shield the window. The arrangement is similar to the familiar spool arrangement used in cameras. By driving the tape slowly past the window during the evaporative cycle, clear vision can be assured, at only the cost of the tape. The drive may be by means of a miniature motor located inside the chamber or by a small magnetic coupling to a motor located outside the chamber.

21.5 *Continuous Systems*

In addition to the systems described, continuous-processing lines are coming into use for high-production items. These units comprise a

Fig. 21.7 Continuous roll coater.

series of chambers separated by doors composed of vacuum gate valves and provided with a mechanical conveyor system which moves carriers containing work assemblies through the series of chambers in a controlled cycle. Such lines may include diffusion ovens, annealing furnaces, and "doping" systems, in addition to conventional vacuum-coating systems. Gas locks are provided between vacuum systems and controlled-atmosphere systems employing hydrogen or inert gases. Such production systems are expensive in first cost and elaborate in operation but are capable of producing great numbers of similar parts at very reasonable costs. Figure 21.7 shows a continuous unit for roll-coating sheet material.

21.6 *Vacuum Furnaces*

Vacuum may be considered an atmosphere just as hydrogen, combusted fuel gas, nitrogen, or argon can be considered an atmosphere. For certain metallurgical operations where extremely pure atmospheres are imperative, no artificial atmosphere, including the noble gases, can approach the degree of purity achieved by even a relatively crude vacuum. To illustrate the point, let us consider that in a furnace where we need extremely high purity of atmosphere we will use argon gas at the best purity level at which it can be purchased, that is, approximately 99.998 percent pure. If the furnace can be completely purged of air and the atmosphere replaced with gas of this purity, which is highly doubtful since complete purging is nearly impossible, we would then have contaminants measured at the rate of 2 parts per 100,000. In actual furnaces, the residual contamination from oxygen absorbed in the brickwork and in the inner surfaces of the retaining shell would probably lead to impurity levels of 10 to 100 times as large as that given. However, if we compare this with the impurity levels in an atmosphere where the pressure has been reduced to 1 micron (1×10^{-3} torr), we will find that the possible contaminants remaining after evacuation will not exceed 1 part in 760,000. This is an improvement in the impurity level by a factor of 7. At a pressure of 1×10^{-6} torr, the total impurities remaining have been reduced 1,000 times, and the total impurity level is therefore 1 in 760,000,000. Of the impurities then remaining, part will be nitrogen, part oxygen, and part presumably water vapor. For most purposes, oxygen is the most important element, and it constitutes only somewhat less than 20 percent of the total residual gas in the vacuum system. We have therefore reduced the active contaminating agent to 1 part in 3,800,000,000. At this level, the total number of oxygen molecules present is usually insufficient to react with even the more sensitive materials to a degree that affects their properties.

The design of high-temperature vacuum equipment for metallurgical purposes involves some knowledge of acceptable materials from the standpoint of their reaction with their environment at high temperature. A reference to the curves for vapor pressure given in Chapter 3 will indicate that many of the materials commonly thought of as suitable for high-temperature work are totally unacceptable for vacuum work. Thus any alloy containing chromium is subject to partial evaporation of the chromium at temperatures of 1600°F and above, which action becomes very rapid at temperatures of 2100°F. The conventional nickel-chromium heating elements and alloy parts are therefore of limited value in vacuum applications. In general, heating elements deteriorate at such a rapid rate that the life of 80-20 nickel-chromium wire or ribbon is measured in hours or days instead of the usual years. On the other hand, relatively thick sections, such as mechanical supports in the form of castings, may be used provided the rough oxidized skin is removed by grinding, machining, or polishing in order to prevent its gas absorption from rendering outgassing difficult or impossible. Such heavy castings do give rise to volatilization of chromium but, being thick, they are usually not weakened sufficiently to be made useless. Ceramic insulating refractory materials can be used in vacuum furnaces but introduce serious outgassing problems because of their pore structure, which readily absorbs large quantities of gas and water vapor. It is therefore usual to limit the use of refractories for furnace insulation to temperatures in the lower 2000°F range. Above this point, insulation is primarily accomplished by means of reflector shields. These are shields made of refractory metals which surround the work area and, being brightly polished, and of course operating in a vacuum, act as quite effective insulators. Shield materials must obviously follow the rules given for temperature, so that they may vary from tungsten through tantalum and molybdenum depending upon the furnace working temperature. In general, the number of shields employed varies from four at 2500°F to as many as nine at 5000°F. There is some question whether shielding with more than six shields will pay its way, since the outer shields accomplish very little. Materials used in shields should grade as one passes through the shielding material from the highly refractory metal on the inside to polished stainless steel on the outside. In cases where this type of insulation is used, the outer furnace shell should be fabricated of stainless steel, water-jacketed, and polished on its inner surface.

Vacuum furnaces are used for brazing operations on heat-resisting alloy materials and superalloy materials such as René 41 as well as for heat-treating or stress-relieving operations on fabrications of these materials. At higher temperatures, such furnaces are used for brazing, stress relieving, or recrystallization of the refractory series of metals and alloys which are used in certain space-vehicle applications and are undergoing intensive study at the present time. Diffusion bonding of titanium and other materials in vacuum is a new and promising joining method now being used for aircraft parts. It is usually carried out in vacuum inside closed retorts.

Vacuum furnaces are also sometimes used for hot-pressing operations on power compacts and for a variety of special laboratory investigations.

21.7 *Vacuum Reduction, Purification, and Refining*

All of the more reactive refractory metals are produced by processes which involve one or more vacuum stages. In general, a salt of the metal is first produced by chemical means, then melted in a closed vessel and some reducing agent, frequently molten magnesium or sodium, added which will reduce the salt of the wanted metal by reacting with it. Thus titanium is produced by reduction of the titanium hexafluoride salt by means of molten magnesium.

At the end of the reduction process, the wanted metal is usually present in the form of a solid sponge, so that the bulk of the reductant product can be drained off through a bottom opening in the vessel. However, in most cases, some of the reducing agent, perhaps magnesium, plus some of its salt, remains behind, held in the particles of the wanted metal. This is frequently removed by a vacuum distillation process, where heat is applied while the vacuum is maintained so that the reducing agent and its salt volatilize and can be removed into a condenser, leaving behind the purified sponge of the wanted metal.

Similar refining operations occur in most of the refractory materials, although the nature of the salts and reducing agents is not identical.

21.8 *Vacuum Casting of Steel*

A great deal of trouble is experienced in massive castings or forgings of alloy steel due to the presence of small residual quantities of hydrogen. This leads to a situation called "hydrogen embrittlement" or "flaking" which can cause failure of large forgings such as turbogenerator shafts under stresses considerably less than their designed loads. Aging at moderate temperatures for lengthy periods is sometimes used to remove the hydrogen, but complete removal in a large forging is still impractical.

The steel industry has recently developed techniques which permit vacuum to be applied to steel, either in a pouring ladle just above the mold into which the metal is to be cast or during the casting process in the mold itself. In either case the vacuum used ranges from approximately 1 micron (1×10^{-3} torr) to approximately 1 torr while the metal is actually in contact with vacuum conditions. If this arrangement is applied while the metal is broken up into a thin stream, as in pouring an ingot or during rapid stirring, the evolution of gas is very rapid and quite complete, resulting in a thorough degassing of the steel, which eliminates the hydrogen embrittlement effects previously mentioned. Figure 21.8 shows one type of steel-degassing system.

Vacuum equipment for this particular operation is relatively crude and is usually accomplished by means of several stages of steam ejectors and interstage heat exchangers which readily handle the large bursts of gas that occur during pouring and are not subject to the wear that would occur in a mechanical vacuum pump were iron oxide allowed to enter.

21.9 *Zone Refining and Crystal Growing*

Semiconductor materials used for electronic devices require extreme purity in their raw material. The purity required is better than can be obtained by the ordinary refining processes used commercially. In addition, these devices generally require that the material be in the form of a single crystal. For this reason, operations called "zone refining" are performed on germanium, silicon, and other materials to increase the purity. The principle is that the material, either self-supported or

Fig. 21.8 Steel-degassing unit. (*Leybold-Heraeus, Inc., Monroeville, Pa.*)

contained in a small boat of some material that does not react with the material being refined, is heated in such a manner that a small molten zone is formed. This is traversed from one end of the bar to the other. In so doing, impurities will migrate in the molten zone, being swept toward the end of the bar, with the result that after several such passes the impurities are concentrated at the ends of the bar. The center is then quite pure material, with impurity levels of a few parts per billion.

Crystal pulling is then carried out utilizing the central portion of the bar for the melt material. In the crystal-pulling operation, a seed crystal of the material being handled, having the orientation desired, is allowed to touch the surface of the molten liquid while the liquid is being maintained within a fraction of a degree of the melting point. If the crystal is then slowly withdrawn, the melt will follow it and slowly freeze upon it so that a carrotlike growth can be achieved by the slow withdrawal of the crystal. Because the process starts with a seed crystal, all the material formed against it will be oriented similarly, and

Fig. 21.9 Ultrahigh-vacuum electron-beam zone refiner: interior details. (*Leybold-Heraeus, Inc., Monroeville, Pa.*)

the full carrot normally is a single giant crystal. This is then sliced up into the necessary number of slices of the proper size for the devices to be made from it.

Both of the operations just described are carried out in vacuum, the exact level depending upon the material being treated. In the case of the refractory metals, no container is known which will permit these materials to be melted without contamination. They are therefore first worked into rod form, and then the rod zone is refined while it is supported at both ends.

Heating for zone refining can be either by means of a single-turn induction coil or by means of an electron beam. The latter type of heating of course requires a pressure level on the order of 1×10^{-4} torr or lower in order that it can operate properly. One such electron-beam zone refiner and its associated vacuum system and controls is shown in Fig. 21.9.

21.10 *Welding*

Certain welding operations on refractory metals and super alloys are best carried out in highly controlled atmospheres. When tungsten–inert gas (TIG) or metal–inert gas (MIG) torches are used with a jet of inert gas passing through the torch, rough control of atmosphere around the melted weld metal is achieved. However, some elements of the air can still reach the weld zone in spite of the blanket of inert gases passing around the weld. If, however, the entire operation is carried out inside a sealed chamber filled with argon or other shield gas, then air can be entirely eliminated. Such chambers are usually purged by being evacuated to 1×10^{-5} torr or lower and are then backfilled with highly purified argon or other shielding gas, so that the welding operation can be carried on at atmospheric pressure but in an atmosphere free of contaminating elements.

For some types of welding work, even such purified atmospheres are not pure enough. This type of work can be carried out by using an electron beam as a heat source at a pressure of 1×10^{-4} to 1×10^{-5} torr without introducing any appreciable contaminants at all. Superior properties can be obtained in some of the refractory metals by this technique. The process provides high welding speeds and extremely narrow heat-affected zones in the joining metals. Welds of aluminum plates as thick as 2 to 3 inches can be accomplished in a single pass under favorable conditions.

21.11 *Electronic Tubes*

The electron-tube industry depends in the last analysis upon the ability to produce a good vacuum inside a sealed-off envelope in order to permit electronic emission and the functions of the tube to be performed. Tubes so diverse in type as oscillograph display tubes, TV picture tubes, and large klystrons for high output of extremely high-frequency energy are similar in their vacuum requirements, although the methods of processing them are quite different. In general, tube evacuation requires the pumping of the tube through some type of glass tubulation while the interior elements of the tube are heated, usually by an induction heater, to a point where outgassing can take place. Simultaneously, the tube envelope is heated in an oven to drive off gas from the envelope. Final evacuation of the tube after sealoff takes place by induction-firing a getter inside the tube to absorb residual gas beyond that which the pump can handle. Getters may be titanium, zirconium, barium, or combinations of these or other metals which have the common property of absorbing gas when heated to a high temperature and deposited as a thin film on the tube envelope. In some tubes residual getters are allowed to remain in the tube at the end of the evacuation and sealoff processes to act as pumps for small amounts of gas which may be liberated during operation of the tube by the bombarding action of electrons passing from the filament to the plate or to the walls of the tube.

Vacuum systems for this work must be of the extremely clean variety, since any back-streaming of oil into the tube would render it inoperative. Internal pressures of 1×10^{-8} torr must usually be maintained during the entire useful life of the tube.

In the mass production of tubes such as are used in home radios and TV's, conveyorized vacuum systems are utilized where the tubes are passed continuously through the various stages enumerated above while pumping is carried on by means of pumps moving with the conveyor.

21.12 *Particle Accelerators*

In connection with physics research, extremely large particle accelerators are being designed and built for accelerating electrons, protons, or alpha particles (helium nuclei) to very high velocities. Such streams of highly energetic particles can be used to examine the interior structure of the atomic nucleus, to transform one species of nucleus to another, or to excite various forms of radiation which may yield information as to atomic structure or binding energies.

The entire accelerating path in such devices must be maintained at a very low pressure, generally on the order of 1×10^{-8} torr or lower. Since some of these circular paths are more than a mile in diameter the problem in achieving leak tightness and complete integrity of the system is obvious. In most cases the system is fabricated of metal and is baked out at some state of evacuation. Rough pumping is accomplished by conventional mechanical pumps which are then completely sealed off from the system by means of bakeable metal seat valves. Final pumping is accomplished by ion pumps or sublimation pumps, which are left on the line for an indefinite period once they are energized. The number of such pumps required for an accelerator such as the Brookhaven or the Stanford lineal accelerator runs into the hundreds, and the control systems for all of these are correspondingly elaborate. Nevertheless such systems are providing a great deal of information about the fundamental particles in physics, which is greatly extending man's mastery over this field. They must in general be treated as special problems, each case being calculated with great care.

21.13 *Laboratory Vacuum Equipment*

A very large variety of vacuum equipment is used in laboratories in connection with an almost endless assortment of hardware to produce special measurements and effects. In some cases use of vacuum is absolutely essential in order to enable the equipment to work, as in the case of an electron microscope, where a carefully columnated beam of electrons is directed through the specimen (or replica thereof for non-electron-transparent materials). The beam is then directed to a photographic plate or to an imaging screen. In the latter case, a coating on the screen fluoresces under the bombardment of the electrons, thus forming a visible image at high magnification. The image on the screen or on the developed film is then viewed through an optical microscope to achieve further magnification. Any such use of an electron beam requires that a good vacuum be held within the system, both to protect the filament from which the electrons are emitted and to avoid dispersion of the beam due to molecular collision between the source and the viewing screen or film.

Variations of this technique include the x-ray microprobe used to excite x-radiation in the material being examined, which may then produce a cathode-ray display, and low-energy electron-defraction equipment used to examine surface atoms in crystals.

Other items of laboratory equipment, such as gravimetric and thermogravimetric balances, employ vacuum not because it is essential to the measurement but to prevent degrading of the data given by the system to an unacceptable level due to random effects or noise. Air must therefore be removed before it can degrade the accuracy so much that it renders the information useless.

Devices such as the vacuum-fusion gas-analysis equipment remove air from the fusion zone in order that gases evolved from the specimen may be collected and kept free from contamination to permit analysis by more conventional absorption or chemical means.

Many measurements of emissivity, reflectance, and other properties of metals and nonmetals must be determined in vacuum, since oxidation of the surfaces during testing might invalidate the result, especially where elevated or moderately hot temperatures are involved.

Spectrometric measurements employing wavelengths shorter than approximately 4000 angstroms must generally be performed in vacuum or partial-vacuum environments in order to avoid attenuation of the incident beam by the air, carbon dioxide, and water molecules present. In addition, the provision of reasonably pure samples usually requires the evacuation of the measuring chamber to a good vacuum followed by admission of the pure gases to be investigated.

In the interest of conserving of space we shall not attempt to list the very large number of laboratory measuring and testing equipments which require the use of vacuum of higher or lower degree in connection with their operation. Suffice it to say that as the measurement of the properties of materials becomes more and more precise, means of eliminating the random noise effects introduced by the atmosphere are becoming ever more necessary, leading to the use of vacuum for more and more of these precise measurements of all kinds. In addition, the investigation of the effects of melting or mixing various materials to give new and better results must usually be carried out in some protective atmosphere or in vacuum in order to eliminate the contaminating effects of air, which could invalidate the conclusions.

21.14 *Vacuum Melting*

In order to purify the refractory metals, especially titanium, now being used in aircraft and space vehicles, single or double vacuum melting is usually necessary. For aircraft-quality titanium, sponge is produced by the reduction of titanium hexafluoride with molten magnesium or sodium. The resulting sponge of titanium metal is washed and purified, ground, and compacted into billets in a high-pressure compacting press. These billets are then welded together, usually by the vacuum electron-beam process, into larger electrodes. Alloy materials are added to the sponge when welding up the electrodes, if desired.

The first melt is of the consumable-electrode type, carried out at a pressure of 10 to 50 microns (1 to 5×10^{-2} torr). This type of electric melting must be used because of the relatively large amount of impurity which still remains from the reduction process. The material is melted into a water-cooled copper crucible. After cooling, the ingot is machined to remove the impurities on the outer surface, and several such ingots are joined together to form a larger ingot for the next melt operation. The second melt is carried out at 1×10^{-4} to 1×10^{-5} torr by the electron-beam consumable-electrode method. The lower pressure allows the removal of much more of the contaminating materials and gases. Thus the characteristics of the final titanium products can be markedly improved by the vacuum melting and purifying operations. Ingots of more than 16 inches diameter, weighing more than 10,000 pounds, have been produced, of a quality suitable for aircraft work. Figure 21.10 shows a small electron-beam melting furnace.

Other refractory metals, and metals of extreme purity requirements, such as tungsten, zirconium, beryllium, tantalum, and alloys of these, are being produced by vacuum processing of a very similar nature. The very high gas loads, the rapid pumpdown required, and the high temperatures involved present very difficult problems to the vacuum engineer.

The consumable electrode melting of high-strength alloy steels is also coming into more widespread use because of the superior fatigue proper-

Fig. 21.10 150-kilowatt electron-beam melting furnace with three indirectly heated guns. (*Leybold-Heraeus, Inc., Monroeville, Pa.*)

ties resulting. Bearings and other fatigue-critical parts of aircraft are now being specified as of vacuum-arc melted material. High strength accompanied by high fatigue resistance is the result.

21.15 *Vacuum-induction Melting*

Vacuum-induction melting, especially of tool steels, is also being used as an alternate method to the consumable-electrode melting of such materials. In this operation, the materials for the heat are assembled in a graphite or silicon carbide crucible and melted by induction from a water-cooled coil surrounding the crucible, powered by rotating or static high-frequency generators. Alloying elements previously placed in hoppers inside the vacuum chamber may be added near the end of the melting period and will be rapidly melted and mixed by the circulation within the melt resulting from the induction-created internal currents. The crucible is then tilted and the melt poured into waiting ingot molds inside the vacuum system. Through the use of internal locks and turntables, repeated charging and removal of ingots can be carried out without release of the vacuum.

Pumping systems usually include a large mechanical pump plus large diffusion pumps or ejector pumps. These must handle not only the large amount of gas from the melted material but also the gas loads incident to the operation of the locks for the admission of melt material and the removal of ingots. This necessitates the use of dual systems to maintain chamber vacuum while the locks are being pumped down, plus an assortment of pneumatically operated valves and controls.

The rapid stirring of the melt resulting from the induction effect results in excellent cleanup of the material, so that the final ingot is not only low in hydrogen and other gases but is also exceptionally clean of nonmetallic inclusions. The result is increased fatigue strength, hardenability, and crack resistivity. This combination of properties does not usually go together in normal steels.

21.16 *Vacuum Casting of Refractory Metals*

As has been mentioned previously, vacuum electron-beam or TIG welding of the refractory metals is possible. Unfortunately, it has not so far been possible to secure high strength and reasonable freedom from brittleness in such welds and in the heat-affected zones immediately adjacent to the welds. As a result, welding is not permissible in highly stressed parts of these sensitive materials.

In an effort to solve these problems, a casting technique has been developed, at first for items required in the atomic-energy program but now extended elsewhere. The process has been used for beryllium, titanium, hafnium, and zirconium and is being extended to tantalum and tungsten as well.

One method consists of using molds machined from pure graphite, cemented together, and preheated and treated in a vacuum furnace. The molds are then placed inside a vacuum-casting chamber in which a crucible of the metal to be cast is placed. Since these materials react with all known crucible materials, the "skull melting" process is used. In this process high-power electron beams are directed at the center of the block or ingot of material to be melted. The effect is to melt only the center of the ingot, the outer portions being maintained in the solid condition by the water-cooled crucible walls. The melting is carried out very rapidly, and when sufficient metal is melted, the crucible is tilted to pour the liquid center portion into the waiting mold.

The outer portion of the crucible contents remains behind, leaving the "skull" which gives the process its name.

A similar process is in use for small and intricate castings, where the mold is produced by the lost-wax process. In this method, a wax replica of the part required is first prepared, then coated with a number of layers of slurry. The first layer is usually of berylia or silica of a high purity, often containing a proportion of tungsten particles. Successive layers of silica slurry are then built up to back up the inner layer. Special proprietary binders, usually containing silicate of soda, are used to harden the slurry layers and to give strength to the mold. The final mold may be backed with molding sand as in ordinary casting, if the pressures require it, but is usually used unbacked. After coating, the mold is heated in an oven to melt out the wax, then assembled inside the skull-melting chamber and cast in a manner similar to the graphite-mold technique.

The graphite-mold method is usually used for larger parts, such as valve bodies, pump casings, actuator devices, etc. The lost-wax mold process is used for such items as pump impellors, valve stems and seats, and structural parts of intricate design for joining aircraft parts, and structural members of titanium.

Both processes produce parts of excellent surface finish, eliminating the necessity for further surface machining on many parts. The lost-wax process can produce small castings with dimensional accuracies within a few thousandths of an inch, which often makes surface machining unnecessary except for the removal of gates and risers, if any.

Where the materials being melted do not react with the crucible materials, induction-melted metal may be used instead of the skull process with either graphite or lost-wax-type molds being used in vacuum melting furnaces. Alloys such as the heat-resisting alloys of nickle, chromium, iron, and molybdenum with tungsten used in turbine blades for jet engines are often cast in this way.

The vacuum systems used for the casting furnaces are of conventional mechanical-pump and diffusion-pump types, but of large capacity, since pressures of 1×10^{-6} or 1×10^{-5} torr are required and must be maintained during the melting and casting cycle in spite of the rapid rate of gas release, as the metal is rapidly melted by the electron beam. Electrical passthroughs are a problem, since beam currents of hundreds of amperes are often used.

The general principles set forth in this book apply to all of these varieties of vacuum equipment to a greater or lesser degree, depending on the degree of vacuum required. Most of the more specialized apparatus is perforce designed by the experimenter who wishes to make use of it, who is not necessarily expert in the fabrication or design of vacuum systems per se. A consideration of the principles set forth herein will enable the construction of the vacuum portion of the experiment to be carried out in a straightforward manner, with the assurance that reliable performance of this part of the apparatus will enable the experimenter to concentrate on the details of his experiment free of unwelcome interruptions due to malfunction of the vacuum portions of his system.

appendix A

Table of Conversion Factors for Pumping System

CONVERSION FACTORS

To convert from	to	multiply by
	PUMPING-SPEED CONVERSION FACTORS	
Cu cm/sec	Liters/sec	0.001
	Liters/min	0.060
	Liters/hr	3.60
	Cu m/hr	0.0036
	Cu ft/min	0.0021
Liters/sec	Cu cm/sec	1,000
	Cu m/hr	3.60
	Cu ft/min	2.12
Liters/min	Cu cm/sec	16.67
	Cu m/hr	0.060
	Cu ft/min	0.0353
Liters/hr	Cu cm/sec	0.278
	Cu m/hr	0.001
	Cu ft/min	0.00059

CONVERSION FACTORS

To convert from	to	multiply by
PUMPING-SPEED CONVERSION FACTORS (*Continued*)		
Cu m/hr	Cu cm/sec	277.8
	Liters/sec	0.2778
	Liters/min	16.67
	Cu ft/min	0.589
Cu ft/min	Cu cm/sec	471.95
	Liters/sec	0.4719
	Liters/min	28.32
	Liters/hr	1,699
LENGTH CONVERSION FACTORS		
Meter	Centimeter	100
	Millimeter	1,000
	Micron	100,000
	Inch	39.37
Centimeter	Millimeter	10
	Inch	0.3937
Millimeter	Meter	0.001
	Centimeter	0.1
	Micron	1,000
	Inch	0.03937
Micron	Meter	0.000001
	Centimeter	0.0001
	Millimeter	0.001
	Inch	0.0000394
Inch	Centimeter	2.54
	Millimeter	25.40
	Micron	25,400
	Meter	0.0254
AREA CONVERSION FACTORS		
Sq cm	Sq in.	0.1550
Sq cm	Sq ft	0.0011
Sq in.	Sq cm	6.452
Sq in.	Sq ft	0.00694
Sq ft	Sq cm	929
Sq ft	Sq in.	144
VOLUME CONVERSION FACTORS		
Cu cm	Liter	0.001
	Cu m	0.000001
	Cu in.	0.061
	Cu ft	0.000035
	U.S. gallon	0.00026

CONVERSION FACTORS

To convert from	to	multiply by
VOLUME CONVERSION FACTORS (*Continued*)		
Liter	Cu cm	1,000
	Cu m	0.001
	Cu in.	61
	Cu ft	0.0353
	U.S. gallon	0.264
Cu m	Liter	1,000
	Cu in.	61,023
	Cu ft	35.3
	U.S. gallon	264
Cu in.	Liter	0.0164
	Cu cm	16.4
	Cu m	0.000016
	Cu ft	0.00058
	U.S. gallon	0.00433
Cu ft	Liter	28.3
	Cu cm	28,320
	Cu m	0.0283
	Cu in.	1,728
	U.S. gallon	7.48

PRESSURE UNITS USED IN HIGH-VACUUM PRACTICE

1 micron (μ) Hg	= 0.001 mm Hg = 10^{-3} mm Hg = 10^{-3} torr
1 micron (μ) Hg	= 1.33 dynes/sq cm
1 millimicron (mμ) Hg	= 0.000001 mm Hg = 10^{-6} mm Hg = 10^{-6} torr
1 millimicron (mμ) Hg	= 1.33×10^{-3} dynes/sq cm
1 millibar (International)	= 0.75 mm Hg = 0.75 torr
1 bar (Bureau)	= 750.00 mm Hg = 29.53 in. Hg = 750 torr
1 mm Hg	= 1,333 dynes/sq cm
	= 1,000 microns = 10^3 microns = 1,000 torr
	= 1.33 millibar
	= 0.0013 bar
	= 0.03937 in. Hg
	= 0.01934 lb/sq in.
750 mm Hg	= 10 dynes/sq cm = 1 megabar
(all at 0°C, and g = 980.6)	
1 in. Hg at 32°F	= 0.4912 lb/sq in.
1 in. Hg at 58.40°F	= 0.4898 lb/sq in.
29.921 in. Hg at 32°F	= 14.696 lb/sq in. = 760 mm Hg = 760 torr
30.000 in. Hg at 58.4°F	= 14.696 lb/sq in. = 762 mm Hg = 762 torr
1 standard atmosphere	= 760 mm Hg = 14.696 lb/sq in. = 760 torr
1 lb/sq in.	= 2.036 in. Hg at 32°F = 51.72 mm Hg = 51.72 torr
1 lb/sq in.	= 2.041 in. Hg at 58.4°F = 51.85 mm Hg = 51.85 torr
1 bar (England)	= 10^6 dynes/sq cm = 750 mm Hg = 750 torr
1 microbar (England)	= 0.00075 mm Hg = 0.75×10^{-3} mm Hg = 0.75×10^{-3} torr

CONVERSION FACTORS

To convert from	to	multiply by
UNITS FOR RATE OF FLOW OF GAS		
1 atm-cu cm/sec	= 1 cu cm/sec at atmospheric pressure	
1 mm-cu ft/min	= 1 cu ft/min at 1 mm Hg (1 torr) pressure	
1 micron-cu ft/min	= 1 cu ft/min at 1 micron (1×10^{-3} torr) pressure	
1 micron-liter/sec	= 1 liter/sec at 1 micron (1×10^{-3} torr) pressure	
1 atm-cu cm/sec	= 760 micron-liter/sec	
	= 1,611 micron-cu ft/min	
	= 1,611 mm-cu ft/min	
1 mm-cu ft/min	= 0.621 atm-cu cm/sec	
	= 1,000 micron-cu ft/min	
	= 472 micron-liter/sec	
1 micron-cu ft/min	= 6.21×10^{-4} atm-cu cm/sec	
	= 10^{-3} mm-cu ft/min	
	= 0.472 micron-liter/sec	
1 micron-liter/sec	= 1.316×10^{-3} atm-cu cm/sec	
	= 2.12 micron-cu ft/min	
	= 2.12×10^{-3} mm-cu ft/min	

appendix B

Bibliography

This book deals primarily in the practical aspects of vacuum systems and devices. Those wishing to dig deeper into the details and theoretical aspects may find the following texts of assistance.

Dushman, Saul: "Scientific Foundations of Vacuum Technique," rev. ed., John Wiley & Sons, Inc., New York, 1962.

Guthrie, Andrew: "Vacuum Technology," John Wiley & Sons, Inc., New York, 1963.

Perani, M., and J. Yarwood: "Principles of Vacuum Engineering," Reinhold Publishing Corporation, New York, 1963.

Roberts, Richard W., and Thomas A. Vanderslice: "Ultra-high Vacuum and Its Application," Prentice-Hall, Inc., Englewood Cliffs, N.J., 1963.

Steinberz, H. A.: "Handbook of High Vacuum Engineering," Reinhold Publishing Corporation, New York, 1963.

Van Atta, C. M.: "Vacuum Technology," McGraw-Hill Book Company, New York, 1965.

Yarwood, J.: "High Vacuum Technique," 3d ed., John Wiley & Sons, Inc., New York, 1955.

Books in related fields:

American Society of Mechanical Engineers: "Pressure Vessel Code," sec. VIII, latest edition.

Bunshaw, R. F.: "Vacuum Metallurgy," Reinhold Publishing Corporation, New York, 1958.

Condon, E. V., and H. Odishaw: "Handbook of Physics," 2d ed., sec. V, McGraw-Hill Book Company, New York, 1967.

Holland, L.: "Vacuum Deposition of Thin Films," John Wiley & Sons, Inc., New York, 1956.

"Industrial Graphite Engineering Handbook," National Carbon Company, Cleveland, Ohio, 1959–1965.

Scott, R. B.: "Cryogenic Engineering," D. Van Nostrand Company, Inc., Princeton, N.J., 1959.

White, Guy K.: "Experimental Techniques in Low Temperature Physics," Clarendon Press, Oxford, 1959.

Symposium transactions dealing with vacuum and space simulation:

American Vacuum Society: symposium transactions 1954–1962, Pergamon Press, New York; symposium transactions 1963, The Macmillan Company, New York. (Series terminated.)

American Vacuum Society: *J. Vacuum Sci. Technol.,* beginning 1967.

American Vacuum Society, Vacuum Metallurgy Division: yearly volumes containing symposium papers.

Institute of Environmental Sciences: symposium transactions beginning about 1960. (Contain increasing amounts of information on space simulation, including both cryogenics and vacuum.)

Review of Scientific Instruments (American Institute of Physics).

Vacuum (Pergamon Press, London).

Index

Index

Accessories, 141–156
bakeout systems, 152–155
expansion joints, 145–146
liquid-nitrogen-level controls, 151–152
passthroughs: electrical, 146–148
liquid and cryogenic, 149
rotary and sliding glands, 149–151
vacuum valves, 142–144
windows, 155–156
Analyzing gas-analyzer readings, 97–98
Avogadro's principle, 176

Backfilling, 160–161
Backing-pump requirements, 33–45, 194
Backstreaming, 35–45
at cutover, 44–45
methods of reduction, 45
in vapor pumps, 35–37
Bakeout systems, 152–155
Bayard-Alpert gauges, 84–85, 115

Bell-jar seals, 139–140
Bibliography, 229–230
Booster pumps, mechanical, 28, 33, 65
Boyle's law, 173, 175
Bryant, Paul, 91

Calibration, vacuum gauge, 100–116
calibration methods in general, 112
comparison method, 101
conductance-limited systems, 109–112
flow-based, 109–111
pressure-based, 111–112
ionization gauges, problems in using, 113–115
Knudsen gauge, 101
McLeod gauge (*see* McLeod gauge)
outgassing of gauges, 115–116

Calibration systems, vacuum gauge, 108–112
flow-based, 109–111
McLeod-based, 108
multiple-conductance-based, 111–112
pressure-based, 111–112
Charles' law, 173–175
Clausing's factor *K*, values of, table, 182
Cleaning systems:
hand and chemical, 158–160
mechanical, 158
Closures, 132–140
bell-jar seals, 139–140
metal-sealed flanges, 137–139
O-ring composite gaskets, 137
O-ring seals, 133–136
pipe threads, 132–133
rectangular O rings, 136–137
Code for unfired pressure vessels, 117
Cold-cathode gauge, 88–92
Cold traps, 6, 34, 41–43
chevron-type, 41–42
effect of, on pumping speed, 43
elbow-type, 42–43
finger-type, 42
Combined gas law, 174–175
Convalux-10 pump fluid, 43
Conversion factors for pumping system, table, 225–228
area, 226
length, 226
pressure units, 227
pumping speed, 225–226
rate of flow, 228
volume, 226–227
Cryogenics in vacuum systems, 67–77
cryogenic arrays, 73–75
surface treatment of, 75
cryopumping calculations, 71–73
lines, 68
liquid-nitrogen systems, 75–76
pumping speed of surfaces, 71–73
shroud materials, 76–77
storage of fluids, 67
use of fluids as pumping mediums, 69
vapor pressure vs. temperature of gases, 70
Cycloidal-type gas analyzers, 95

DC-704 and DC-705 pump fluids, 43
Degassing methods, 19–22
bakeout, 19
cryogenics, 21
ion bombardment, 22
time effects, 22
Degradation effects in vacuum, 20–21
cryogenic effects, 21
organics and plastics, 20
time effects, 21
Diffusion-pumped vacuum systems, 4–6

Electrical passthroughs, 146–148
Expansion joints, 145
Expansion ratios, 3
Extractor gauge, 92

Finishing, cleaning, and backfilling systems, 157–161
backfilling, 160–161
hand and chemical cleaning, 158–160
mechanical cleaning, 158
Flow of gases (*see* Gases, flow of)
Fluids and plastics, vapor pressures of, 9-12
pump fluids (*see* Vapor-pump fluids)

Gas analyzers, vacuum (VGAs), 94–99
analyzing the readings of, 97–98
cycloidal-type, 95
monopole-type, 96
omegatron-type, 95–96
problems in using, 97–98
access, 97
bakeout, 97
stability, 97
quadrupole-type, 96
sector-type, 94–95
specifying characteristics of, 98–99
resolving power, 98
sensitivity, 99
unit resolution, 98
time-of-flight type of, 96
Gas constant, universal, 176–177
table, 177
Gases:
flow of, 179–189

Gases, flow of (*Cont.*):
flow conductance and impedance, 183–185
importance of computing net pumping speed, 187
mean free path, 181
molecular flow, 181–183
Clausing's factor *K*, values of, table, 182
pumpdown time, 185–189
effect of mixtures of gases, 188–189
molecular flow, 185–186
viscous flow, 185–187
viscous flow, 179–181
molecular dimensions of, 163
temperature vs. vapor pressure, 70
theory of, 171–178
Avogadro's principle, 176
Boyle's law, 173, 175
Charles' or Gay-Lussac's law, 173–175
combined gas law, 174–175
gas constant, 176–177
table, 177
nature of gases, 172
states of matter, 171–172
velocity of gas molecules, 177–178
Gauges, vacuum, 78–93
Bayard-Alpert, 83–86
cold-cathode, 87–92
extractor, 92
inverted magnetron, cold-cathode, 88–92
ionization, 79, 81–83
cold-cathode, 87–88
Nottingham, 86–87
Pirani, 81
readout devices, 92–93
relative sensitivity for various gases, 83
thermal-response, 79
errors of, 81
thermocouple, 79 81
triggered Penning, 90–91
errors of, 91
triode, 82
types of, 78–79
absolute-pressure, 78
ionization, 79
partial-pressure, 79

Gauges, vacuum, types of (*Cont.*):
thermal-response, 79
use of, 6–113
location, 6
problems in, 113
Gay-Lussac's law, 173–174

Hall, Lewis D., 46, 49
Hastings gauge, 80
Helium-leak detectors, 164–167
Helmer and Hayward gauge, 92
Hobson, J. P., 89

Inverted magnetron gauge, cold-cathode, 88–92
Ion-pumped systems, 6–7
Ion pumps, 46–55
cleaning and repairing, 62–64
Ionization gauges, 79, 81–87
cold-cathode, 87–88

Jepson, Robert L., 46, 48

Knudsen gauge, 63, 101
Kreisman gauge, 63, 89

Leak detection, 162–170
gas molecular dimensions, 163
helium-leak detectors, 164–167
necessity for, 162–163
repairing leaks, 167
total-leakage measurements, 168–170
troubleshooting, 167–168
using acetone and gauges, 163
Liquid and cryogenic passthroughs, 149
Liquid-nitrogen-level controls, 151–152

McLeod gauge, 101–109, 112
accuracies of, calibration, 108–109
derivation of formula for, 103–104
errors of: condensibility, problem of, 106
Guidé-Ishii effect, 105–106
inaccuracy of feducial mark, 105

McLeod gauge, errors of (*Cont.*):
long-pumpdown effects, 106
static charges, 104–105
sticking effects, 104
temperature changes, 105
use of, for calibration purposes, 106–108
Materials of construction, 8–22
absorption effects on, 18
cryogenic effects on, 21
degassing chambers, alternate methods of, 22
degradation effects on, 20–21
metallic oxides, effects on, 15–17
metals, bakeout effects, 18–20
plastics and organic materials, 20
time effects on, 21–22
vapor-pressure information, importance of, 8
vapor pressures: of metals, 12–15
of organic materials, 9
table, 10
of pump fluids and plastics, 9, 12
tables, 11, 12
Mean free path, 181
Metal-sealed flanges, 137–139
Metals, 12–14
bakeout effects, 18–20
oxides, 15–17
Micron, 2
Molecular-drag pumps, 29–30
Molecular flow, 181–183
Monopole-type gas analyzers, 96

Nottingham gauge, 86–87

O ring, 133–137
composite gaskets, 133–136
seals, 133–136
rectangular, 136–137
Omegatron-type gas analyzers, 95–96
Organic materials, 9
table, 10

Passthroughs:
electrical, 146–148
liquid and cryogenic, 149
Penning gauge, 87–89
Pipe threads, 132–133
Pirani gauge, 81, 88
Plastics and fluids, vapor pressures of, 9–12
Pump fluids (*see* Vapor-pump fluids)
Pumpdown time, 185–186
Pumping calculations, 190–201
calculations for system with metal and plastic parts, 195–198
outgassing, 196–198
roughing time, 195
time to reach 10^{-6} torr, 196
designing an ultrahigh-vacuum system, 198–201
general requirements, 198–199
handling load outgassing problems, 200–201
total pumpdown time, 200
selection of pumps for bell-jar coater, 190–194
backing-pump requirements, 194
diffusion pump and connecting piping, 192–194
low-pressure piping, 191
mechanical pump, 191
Pumps, vacuum: application of, 5–7
diffusion, 5
ionic, 6
Roots type, 7
sorption, 7
turbine, 7
cleaning and repairing, 56–66
diffusion pumps, 59–62
assembling, 61
oil breakdown or contamination, 59
tar deposits, 59–60
ion-pump elements, 64–65
ionic pumps, 62–64
cleaning elements, 64
starting problems, 63
mechanical booster and turbine pumps, 65–66
mechanical pumps, 57–59
contaminated oil, 57–58
mechanical difficulties, 58–59
description of, 23–55
ionic, 46–55
determination of pump speed by two-gauge method, 51–52
method of construction, 47–48

Pumps, vacuum, ionic (*Cont.*):
pumping mechanisms, 46–49
use of pump as a gauge, 54
mechanical, 23–31
mechanical booster (Roots), 28
molecular-drag or turbine, 29–30
purging, 27–28
rotary-type, 25–28
vane-type, 23–25
water-sealed, 31
operating life, 50
relative speeds for ion pump and sublimator, 53
reliability, 65
repairing (*see* cleaning and repairing *above*)
selection of type of, 54–55
sublimation types, 49–50
filament life, 50
vapor-type, 32–40
diffusion, 33–37
speeds, 37–40
ejector, oil and steam, 32–33

Quadrupole gas analyzers, 96

Readout devices, vacuum gauge, 92–93
Redhead, Paul A., 89
Redhead gauge, 89, 91
Repairing leaks, 167
Roots blower, 7, 28
Rosenberg, Paul, 102
Rotary and sliding glands, 149–151
Rotary-type pumps, 25–28

Santeler, Donald J., 74
Scheumann gauge, 92
Sector-type gas analyzer, 94–95
Shrouds, 73–77
materials, 76–77
types of, 73–77

Theory of gases (*see* Gases, theory of)
Thermal-response gauges, 79
errors of, 81
Thermocouple gauges, 79–81
Time-of-flight gas analyzer, 96
Torr, definition of, 2
Total-leakage measurements, 168–170
Trapping systems (*see* Cold traps)
Triggered Penning gauge, 90–91
Triode gauge, 82
Turbine pumps, 29–30

Unfired Pressure Vessel Code, 117
Uses of vacuum systems, 202–224
crystal growing, 215–216
electronic circuit, 207–211
continuous systems, 210–211
evaporating, 207–209
sputtering, 209–210
electronic tubes, 217
furnaces, 211–213
laboratory equipment, 218–220
melting, 214–223
casting of refractory metals, 222–223
consumable electrode, 224
induction, 221
"skull melting" and casting, 222
steel casting, 214
metallizing equipment, 202–207
particle accelerators, 218
reduction, refining, and purification, 213
welding, 216–217
zone refining, 214–215

Vacuum gas analyzers (*see* Gas analyzers, vacuum)
Vacuum gauge calibration (*see* Calibration; Calibration systems)
Vacuum gauges (*see* Gauges, vacuum)
Vacuum pumps (*see* Pumps, vacuum)
Vacuum systems:
definitions of, 1
degrees of vacuum, 2
required parts for, 4
uses of (*see* Uses of vacuum systems)
Vacuum vessels (*see* Vessels, vacuum)
Valves, vacuum, 142–144
Vane-type pumps, 23–25
Vapor pressure of materials, 8–20
adsorption effects, 18–20
on metals, bakeout effects, 18–20
on plastics, 20

Vapor pressure of materials (*Cont.*):
fluids and plastics, 9–12
importance of, 8
metallic oxides, 15–17
metals, 12–15
organics and plastics, 9
Vapor-pump fluids, 12, 43
characteristics of, table, 12
Convalux-10, 43
DC-704 and DC-705, 43
Vessels, vacuum, 117–123
aluminum, 121
carbon steel, 120–121
copper and brass, 119–120
glass, 118–119
mechanical considerations, 117–118
stainless steel, 122–123
Viscous flow, 179–180
Viton-A, 12

Welding, 124–131, 216–217
for high vacuum, 124–131
general requirements, 124–126
metal–inert-gas (MIG), 125
tungsten–inert-gas (TIG), 122, 125, 126, 128–129
weld preparations and techniques, 126–127, 129–130
welder qualifications, 127–128
welding rods, 126
for ultrahigh-vacuum systems, 130–131
Wilson seal, 149–150
Windows, viewing, for vacuum chambers, 155–156

Young, J. R., 90, 91